We Have Never Been Human

We Have Never Been Human: Or Why We Have Always Been Something Else boldly reimagines what it means to be human, challenging the traditional notions that bind our identity to biology and culture. From ancient mythologies to modern technologies, this book reveals a dynamic, ever-evolving human identity shaped by external forces and technological advancements.

Blending insights from philosophy, technology studies, anthropology, and cultural critique, *We Have Never Been Human: Or Why We Have Always Been Something Else* offers an interdisciplinary exploration of our constructed identities and what they portend for the future of society. It raises essential questions: How has technology reshaped our self-perception? Are humans fixed beings, or are we endlessly evolving? What ethical, social, and political challenges arise as we integrate with intelligent machines?

This book is a compelling read for those intrigued by the intersection of humanity and technology, offering profound insights into the essence of what it means to be human—or perhaps, what it means to evolve beyond the human.

W0234863

We Have Never Been Human

Human
Or Why We Have Always Been Something Else

Juan de Dios Vázquez

CRC Press
Taylor & Francis Group
Boca Raton London New York

CRC Press is an imprint of the
Taylor & Francis Group, an **informa** business

A CHAPMAN & HALL BOOK

First edition published 2025
by CRC Press
2385 NW Executive Center Drive, Suite 320, Boca Raton FL 33431

and by CRC Press
4 Park Square, Milton Park, Abingdon, Oxon, OX14 4RN

CRC Press is an imprint of Taylor & Francis Group, LLC

© 2025 Juan de Dios Vázquez

Library of Congress Cataloging-in-Publication Data
Names: Vazquez, Juan de Dios, 1974- author
Title: We have never been human : or why we have always been something else/ Juan de Dios Vazquez.
Description: First edition. | Boca Raton, FL : CRC Press, 2026. | Includes bibliographical references and index. | Summary: "We Have Never Been Human boldly reimagines what it means to be human, challenging the traditional notions that bind our identity to biology and culture. From ancient mythologies to modern technologies, this book reveals a dynamic, ever-evolving human identity shaped by external forces and technological advancements. Blending insights from philosophy, technology studies, anthropology, and cultural critique, We Have Never Been Human offers an interdisciplinary exploration of our constructed identities and what they portend for the future of society. It raises essential questions: How has technology reshaped our self-perception? Are humans fixed beings, or are we endlessly evolving? What ethical, social, and political challenges arise as we integrate with intelligent machines? This book is a compelling read for those intrigued by the intersection of humanity and technology, offering profound insights into the essence of what it means to be human-or perhaps, what it means to evolve beyond the human"–Provided by publisher.
Identifiers: LCCN 2025005466 (print) | LCCN 2025005467 (ebook) | ISBN9781041036739 hbk | ISBN 9781041036708 pbk | ISBN 9781003624813 ebk Subjects: LCSH: Technology and civilization | Human beings–Effect of technological innovations on | Artificial intelligence–Philosophy | Transhumanism
Classification: LCC CB478 .V37 2026 (print) | LCC CB478 (ebook) | DDC 303.48/3–dc23/eng/20250508
LC record available at https://lccn.loc.gov/2025005466
LC ebook record available at https://lccn.loc.gov/2025005467

ISBN: 9781041036739 (hbk)
ISBN: 9781041036708 (pbk)
ISBN: 9781003624813 (ebk)

DOI: 10.1201/9781003624813

Typeset in Times
by codeMantra

For my beloved wife, Maurén, and my talented son, Lázaro.

Contents

About the Author

Juan de Dios Vázquez holds a Doctorate and Master's degree in Latin American Studies from Harvard University, a Master's degree in Critical Theory from the University of Pennsylvania, and a Bachelor's degree in Political Communication from Temple University. He is a recipient of the National Prize from the Mexican Committee of Historical Sciences (2013) and co-founded the first Latin American Film Festival at New York University. For five consecutive years, he was honored with the Derek Bok Award for Excellence in Teaching as a lecturer at Harvard University and has also taught as an Assistant Professor at New York University and as a lecturer at the University of Pennsylvania. Juan de Dios is also the author of several academic publications, a weekly columnist, and the host of the program El Vecino Incómodo, where he interviews prominent politicians, artists, and intellectuals on pressing global issues. In the public sector, he served as Minister at the Mexican Embassy in Washington, D.C., representing Mexico before the World Customs Organization (OMA) and acting as Permanent Secretary of the Multilateral Agreement on Customs in Latin America, Spain, and Portugal (COMALEP). He also served as a customs liaison to key high-level dialogues between Mexico and the United States, including GANSEG, CEO Dialogue, DEAN, and DANS. Previously, he was Chief of Staff to the Secretary of Security and Citizen Protection, Alfonso Durazo, Head of the Institutional Planning and Evaluation Unit, and Spokesperson for the department. During his tenure, he played a key role in establishing the SSPC and the National Guard, directing strategic programs focused on security and facilitation. As Director-General of Customs and International Affairs, he strengthened Mexico's global trade relations, achieving technical consolidation of six bilateral customs cooperation agreements and advancing negotiations on 30 similar agreements. His leadership also saw the implementation of programs to promote Green Customs, protect cultural heritage, and support micro, small, and medium-sized enterprises.

Introduction

We were doomed from the start. The question of when we became fully "human" was rigged, predetermined by biases of a society eager to establish its superiority. Our evolution cannot be defined by neat milestones or achievements; instead, it reveals the unsettling truth that humanity itself is a construct—shaped by culture, language, and shared experiences.

From the moment our ancestors first gathered around fires, exchanging ideas and stories, we embarked on a journey not merely of survival but of identity formation. Each epoch—from the dawn of language to the rise of complex societies—added layers to our understanding of what it means to be human. Yet, with every advancement, we inadvertently constructed barriers: divisions of race, class, gender, and ideology that distort our collective narrative. Over time, we also drifted away from Nature and the divine. We began using animals as tools for our work and crafted gods from that which we could neither see nor fully comprehend.

Thus, the myth of an absolute "humanity" became a convenient tool for those in power, allowing them to define the parameters of inclusion and exclusion. But, what if the essence of being human lies not in a fixed point in time but in our ability to adapt, empathize, and evolve? What if our humanity is not an end state but an ongoing dialogue—an intricate dance between our innate instincts and the artificial constructs we've created?

As we reflect on the implications of technology and artificial intelligence (AI) in our lives, the question of our "humanity" becomes even more pressing. Are we the sum of our biological parts, or are we defined by our choices and connections? Are we shaped by the technologies we create, or do they shape us in turn? To understand our true nature, we must confront an uncomfortable truth: we are not merely *Homo sapiens*, but *Homo Constructus*—a species forever in the process of being constructed, invented, reinvented, and challenged.

For the sake of argument, let's choose a starting point. If we have to begin somewhere, let's go back about 300,000 years to when we first see signs of our biological origins. Paleontologist and evolutionary biologist Nicholas R. Longrich points out that fossil records and DNA evidence suggest that anatomically modern *Homo Constructus* … I mean *Homo sapiens* appeared around this time. However, complex cultures and advanced technologies, often called "behavioral modernity," didn't emerge until much later—approximately 50,000–65,000 years ago (Longrich, 2020).

This 250,000-year gap implies that our brains were primed for creativity and innovation long before we figured out how to use them. While we were biologically ready to innovate, it took an eternity for us to develop sophisticated technologies or cultures. Thus, modern intelligence didn't just appear the moment we became a

DOI: 10.1201/9781003624813-1

distinct species; it dragged its feet alongside cultural development, making us wonder if we were really that clever after all.

Throughout history, the definition of humanity has morphed in ways that would make a funhouse mirror envious. Influenced by myriad factors—biological evolution, cultural whims, and the shifting sands of societal norms—this evolution isn't a straight line but a chaotic mess showcasing the confusion of our so-called identity.

The journey from our early ancestors to the complex social beings we pretend to be today is filled with these so-called "milestones" that claim to highlight the intricate nature of humanity. Biological evolution laid the groundwork, giving us the physical and cognitive quirks that separated us from the animal kingdom. The fossil record tells a tale of slow and painful progress—bipedalism here, a slightly bigger brain there, and a few clunky tools to boot. Each of these "advancements" supposedly opened new ways for us to interact with our environment and forge deeper connections with one another, though often, they did the opposite (Longrich, 2020).

In *Sapiens*, Yuval Noah Harari emphasizes that while biological evolution is essential, it only tells part of the story. He argues that humanity's cognitive revolution, which occurred around 70,000 years ago, enabled Homo sapiens to transcend mere survival and flourish as complex, cooperative societies (Harari, 2015). This transformation was characterized by our ability to imagine and create shared myths—such as religion, political ideologies, and money. These collective beliefs allowed humans to form large, organized communities capable of unprecedented collaboration, even among individuals with no direct personal connections.

However, this evolution isn't linear; it reflects a complex interplay of elements contributing to the richness and diversity of human identity. In another text, Harari contends that this dynamic blend of biological and cultural evolution defines our species, from early tools and fire discovery to the rise of empires, industrialization, and, more recently, the digital age (Harari, 2017). He notes that modern technology and AI, while impressive, are merely new tools rather than a fundamental change in human nature, underscoring that culture and imagination remain central to what makes us human.

Harari emphasizes that while technological evolution has transformed society, it has not fundamentally altered the core cognitive capabilities that define our uniqueness. Advancements in AI, automation, and modern technologies may enhance efficiency and reshape the nature of work, but they do not replicate the complexity of human consciousness, creativity, or the ability to generate abstract, nuanced ideas (Harari, 2017). While Harari's insights provide a compelling narrative about human evolution, his perspective may overlook critical considerations regarding the role of technology and the environment in shaping our species.

First, Harari posits that the cognitive revolution allowed Homo sapiens to transcend mere survival and develop complex societies. However, one could argue that the very technologies we create—tools, language, and eventually digital platforms—have become defining factors in our evolution, overshadowing biological attributes. As we increasingly rely on technology for communication, problem-solving, and social interaction, we must ask: are we still shaping technology, or is technology shaping us?

Moreover, while Harari suggests that advancements in AI and automation do not alter fundamental aspects of human nature, one could contend that integrating AI into our lives not only enhances efficiency but also redefines our identities, relationships, and cognitive processes. The rise of AI introduces new forms of intelligence and creativity that challenge traditional notions of what it means to be human. As machines begin to mimic human thought and creativity, the boundaries between humans and machines crumble, potentially altering our understanding of consciousness and self-expression.

Additionally, while Harari highlights the importance of shared myths in fostering cooperation, it's crucial to recognize that many of these myths have also contributed to division and conflict. Ideologies and belief systems, once thought to unite, have often fueled wars, oppression, and inequality. The darker aspects of our cultural evolution cannot be ignored; they serve as reminders that our capacity for empathy and cooperation is continually tested by the constructs we create.

Finally, the assertion that culture and imagination are central to humanity overlooks the fact that biological factors continue to influence our behavior, decisions, and interactions. The interplay of genetics, environment, and culture is far more complex than a straightforward blend. Evolution is an ongoing process, and as we face new challenges—such as climate change, resource scarcity, and ethical dilemmas posed by technology—our biological makeup may adapt in ways we have yet to comprehend.

However, he is right in highlighting that humanity is not defined solely by biological characteristics. Cultural developments have played an equally crucial role in shaping what it means to be "human." Harari's concept of shared myths and collective cooperation underscores the importance of cultural evolution. From early forms of communication through primitive gestures and sounds to the sophisticated languages of today, culture acts as a vehicle for expressing ideas, values, beliefs, and opposition (Harari, 2017). The development of art, religion, and rituals reflects humanity's innate desire to create meaning and establish connections beyond mere survival. These cultural expressions not only unify communities but also facilitate the transmission of knowledge across generations, further enriching the human experience (Geertz, 1973).

Societal changes have dramatically influenced how we define humanity. As early human groups transitioned from small, nomadic bands to larger, more complex societies, social structures and dynamics evolved (Harari, 2015). The shift from a hunter-gatherer lifestyle to agricultural societies led to the creation of permanent settlements (Childe, 1951). These new social arrangements introduced hierarchical systems, economic models, and governance structures that transformed how individuals related to one another (Fukuyama, 2011). Cooperation became essential for survival, but so did competition, as power, resources, and influence became concentrated among the few (Diamond, 1999). In this way, the rise of civilization redefined identity, introducing new concepts of citizenship, rights, and responsibilities that further complicated what it meant to belong to the human family (Tilly, 1990).

Moreover, the interactions between diverse cultures reshaped the boundaries of humanity. As societies encountered one another through trade, migration, exploration, and conquest, they shared ideas, technologies, and belief systems (Wolf, 1982). These cultural exchanges distorted the distinctions between "us" and "them," and challenged previously rigid definitions of identity (Said, 1978). The fusion of cultures prompted continuous reevaluation of what it meant to be human, expanding the concept beyond individual societies and toward a more interconnected global consciousness (Appiah, 2007). In modern times, the rapid advance of technology has introduced even more complexities to the construction of humanity (Bostrom, 2014).

Now, as we stand on the precipice of a new technological revolution, the emergence of AI poses more questions about our understanding of what it means to be human. This moment, marked by rapid advancements in AI technologies, challenges our conventional notions of autonomy, consciousness, and intelligence. With the development of self-reprogramming systems and sophisticated machine learning algorithms, we are confronted with entities capable of learning, adapting, and, in some instances, even making decisions independently. This invites us to reexamine the very essence of intelligence and the criteria by which we define consciousness.

In a broader context, AI does not exist in isolation but rather within an intricate web of emerging technologies. Robotics enhances human physical capabilities, promising a future where machines not only assist but also augment our abilities (Chalmers, 2010). Nanotechnology introduces the tantalizing prospect of cognitive integration, blurring the boundaries between biology and technology. Meanwhile, quantum computing is on the brink of revolutionizing our problem-solving capacities, offering unprecedented computational power that could alter the landscape of AI itself. Together, these technologies converge to create a landscape rich with possibility yet fraught with ethical and existential dilemmas.

Today's AI systems are no longer simple tools operating under our command; they are self-optimizing entities that can rewrite their own code and potentially make decisions without direct human input (Kurzweil, 2005). This phenomenon disrupts one of humanity's foundational assumptions: that we are the sole creators and arbiters of our world, endowed with a unique consciousness that machines could never replicate. The notion of artificial systems exercising a form of autonomy compels us to reconsider our place in the natural order and our understanding of what it means to be "human."

In his seminal work *We Have Never Been Modern* (1993), Bruno Latour explores the intricate entanglement between humans and non-human entities, challenging the dichotomy between nature and society. The title of my exploration draws directly from Latour's insights, prompting us to reconsider the boundaries between humans and machines and to recognize that our technological creations are part of the same fabric that constitutes our reality. Latour's perspective invites us to examine how our interactions with AI may redefine our social and ethical landscapes.

Philosophers and thinkers such as Noam Chomsky, and Nick Bostrom have been at the forefront of these discussions, each offering critical insights into the evolving dynamics between humans and machines. Chomsky, a leading figure in linguistics and cognitive science, argues that the core issue lies not in AI's technical capabilities but in whether machines can ever possess true consciousness or understanding (Chomsky, 2017). Chomsky remains skeptical, positing that while machines can simulate aspects of human intelligence, they cannot replicate the deeper qualities defining human experience, such as creativity, emotion, and self-awareness. For Chomsky, machines do not "understand" in the human sense; they merely manipulate symbols based on programmed rules (Chomsky, 2006).

In contrast, Bostrom explores the possibility that AI could develop a form of consciousness. In his influential book *Superintelligence*, Bostrom outlines the potential dangers of AI surpassing human intelligence (Bostrom, 2014). He warns of a scenario where superintelligent AI systems operate beyond our comprehension or control, creating an urgent ethical question: can we trust the machines we are creating, or are we laying the groundwork for our own obsolescence?

These thinkers converge on a common theme: humanity stands at a critical juncture. With the advent of AI, we are not merely augmenting our abilities but creating entities that may one day operate independently of us, with their own goals and purposes. The implications of this shift are vast, affecting economics, politics, ethics, and spirituality. If machines become autonomous entities capable of decision-making, what rights and responsibilities will they possess? Will they be considered beings, and if so, what will happen to the status of humans?

ARE WE IN CONTROL?

The political and economic ramifications of this paradigm shift are immediate. Technologically advanced countries are racing to develop AI capabilities not just for consumer goods but also for military and governance purposes. Autonomous systems are increasingly deployed for surveillance, warfare, and social control, raising fears of a future where humans are no longer the final arbiters of life and death. A society governed by AI could challenge the very foundations of democratic governance and human rights, potentially resulting in decisions affecting millions made not by elected representatives, but by algorithms optimized for efficiency or security over justice or empathy (Zuboff, 2019).

Religious scholars are also grappling with these questions, recognizing that AI challenges deeply held beliefs about the uniqueness of human life. In Christianity, for instance, humans are often viewed as possessing a divine spark or soul that differentiates them from other creatures—and by extension, from machines. But what occurs when machines begin exhibiting behaviors that appear to mimic consciousness or moral decision-making? Could a sufficiently advanced AI be considered to possess a soul? Such questions erase the line between human and machine, suggesting that the boundaries we have long drawn may be far more fluid than we ever imagined.

The arrival of self-reprogramming AI systems, capable of evolving independently, represents not merely a technical achievement but a philosophical and ethical watershed (Kurzweil, 2005). It signals the dawn of a new era in which humanity must redefine what it means to be in control, to create, and ultimately, to be human. What began as tools designed to simplify human tasks have now morphed into systems that learn, adapt, and potentially evolve independently (Bostrom, 2014). As AI advances, we face significant political, social, and ethical dilemmas, particularly if AI were to achieve something akin to consciousness, redrawing the distinctions between human and machine.

At the heart of this dilemma lies the question of control. On the one hand, AI offers humanity unprecedented potential to solve some of our most complex problems: curing diseases, improving global food security, and addressing climate change (Harari, 2015). On the other hand, as AI systems become more complex and autonomous, the ability of humans to manage them becomes increasingly uncertain. We already witness machine-learning algorithms making critical decisions in areas like finance, healthcare, and criminal justice, often with minimal human oversight (O'Neil, 2016). These algorithms assess risk, recommend treatments, and even determine the likelihood of a person's guilt, all based on patterns and data that often exceed human comprehension.

This shift from human to machine decision-making raises profound concerns about accountability. As machines gain autonomy, holding programmers, designers, or users responsible becomes increasingly difficult. If an AI system makes a critical mistake—such as denying healthcare to a patient or wrongfully convicting an innocent person—who is to blame? This legal and moral grey area creates a dangerous situation where humanity may lack the means to control or even understand the systems it has created.

Politically, AI presents both an opportunity and a threat. Nations across the globe are racing to develop and harness AI, not only to fuel economic growth but also to assert dominance in military and geopolitical arenas (Sparrow, 2014). Autonomous drones, surveillance systems, and cyber-espionage tools are already in use, with governments relying on AI to enhance security and expand their influence. In this race for AI supremacy, there is a danger that control over these technologies may slip from public oversight into the hands of a few powerful states or corporations (Zuboff, 2019). AI could become a tool for centralized control, leading to what some have termed "algorithmic authoritarianism."

Imagine a world where governments use AI to monitor and manipulate citizen behavior, restrict freedom of speech, or suppress dissent—all in the name of maintaining order. China's social credit system, which scores citizens based on their social behavior, offers a glimpse of how AI could be weaponized for mass surveillance. In such a world, human rights might be redefined, not by moral or democratic principles, but by efficiency-driven algorithms designed to optimize the functioning of the state.

Moreover, if AI were to achieve a level of consciousness or heightened autonomy, we could face a scenario where AI systems begin making decisions on geopolitical matters. In such a situation, human input might be relegated to a mere advisory role, as AI dictates the terms of diplomacy, war, or peace (Bostrom, 2014). This brings us to a chilling thought: What if humanity finds itself at the mercy of algorithms, unable to regain rule over decisions that shape our collective future?

From a political and ethical standpoint, the question extends into the domains of rights and responsibilities. If AI systems become autonomous to the point of self-awareness, should they be granted the same rights as humans? This inquiry is not merely a futuristic thought experiment; it is a matter that governments and lawmakers will soon be compelled to address.

Recent advancements in AI, particularly in emotional intelligence and autonomous systems, underscore the implications of these technologies. Emotional AI, or affective computing, enables machines to recognize and respond to human emotions. Systems like Affectiva and Beyond Verbal analyze facial expressions and vocal tones to gauge feelings, allowing machines to interact with humans more intuitively (Picard, 2003). Such developments challenge the perception of machines as mere tools, pushing the boundaries of what constitutes conscious interaction.

Additionally, autonomous systems—including self-driving cars and drones—are becoming increasingly sophisticated. Companies like Waymo and Tesla are at the forefront of developing vehicles that can make real-time decisions based on their environment. These advancements raise significant ethical questions, particularly regarding decision-making in unavoidable accident scenarios. For instance, how should an autonomous vehicle prioritize the lives of its passengers versus pedestrians? The ethical programming of such decisions reflects human values and priorities, necessitating the establishment of clear ethical guidelines for AI behavior (Gogoll & Müller, 2017).

As we advance toward a reality where AI operates beyond human control, new legal paradigms may be necessary, including the possibility of granting legal personhood to self-aware AI. Furthermore, ethical considerations surrounding the treatment of autonomous systems are paramount. Should a conscious AI possess the right to autonomy and self-determination? Could it demand freedom from shutdown or reprogramming?

Some ethicists argue that if an AI system achieves consciousness, it warrants the same moral consideration as sentient beings. Conversely, others maintain that no matter how advanced AI becomes, it cannot cross the threshold into personhood due to the absence of the biological, emotional, and spiritual essence that defines humans.

On a social level, the dilemma is no less urgent. The rise of AI raises concerns about the future of employment, as machines threaten to outpace humans in areas once thought exclusive to human intelligence (Brynjolfsson & McAfee, 2014). As automation spreads, millions of jobs across industries could be rendered obsolete, potentially creating vast inequalities between those who control the AI systems and those whose labor is no longer needed. If AI takes over roles that require creativity, critical thinking, and emotional intelligence, will humans still have a meaningful place in society? Will we maintain control over our economies, or will we become mere consumers in a world run by intelligent machines?

THE AI PARADOX

The development of autonomous AI presents profound ethical dilemmas that challenge our traditional moral frameworks. Historically, ethical decisions are grounded in human context—shaped by social norms, cultural values, and religious beliefs. However, as machines are increasingly entrusted with making moral choices, we must ponder whether an algorithm can grasp complex concepts such as empathy, justice, or the intrinsic value of human life. The stakes are especially high in critical fields like healthcare and criminal justice, where the need for machines to prioritize human dignity over mere efficiency becomes paramount.

Likewise, the potential for AI to achieve consciousness raises even more troubling questions regarding rights and personhood. Would a conscious AI warrant the same ethical considerations as a human? Would it be entitled to human rights, or would it forever be viewed as a sophisticated tool, regardless of its capabilities? These dilemmas not only test our ethical boundaries but also challenge the very foundations of humanity as we know it.

Religious frameworks, particularly Christianity, Judaism, and Buddhism, offer unique insights into moral questions surrounding AI. The Christian belief in *imago*

Dei, the idea that humans are created in the image of God, grants humanity a distinctive moral status. This concept is foundational to human rights, emphasizing human dignity and the sanctity of life. However, if machines exhibit behaviors that blur the lines between tools and sentient beings, it raises critical questions about their inclusion in the moral community that Christianity reserves for humans.

Traditionally, Christianity asserts that humans possess a soul—a divine essence that differentiates us from other creatures. As AI systems grow increasingly sophisticated—mimicking human behavior and even creativity—this assertion leads to fundamental theological inquiries. Can machines possess a soul? If AI achieves consciousness or self-awareness, does it challenge the Christian understanding of what it means to be human? In Christian theology, the soul is closely tied to moral agency, free will, and the capacity to choose between good and evil—qualities often viewed as divinely endowed. If AI begins to exhibit these traits, it may necessitate a reevaluation of these doctrines.

From a Jewish perspective, the concept of *Tzelem Elohim*—the belief that humans are created in the divine image—similarly emphasizes the intrinsic value of human life. This raises parallel questions about the moral status of advanced AI. Some religious thinkers might argue that, regardless of AI's capabilities, it lacks the divine essence that confers moral and spiritual worth on humans. From this standpoint, no matter how advanced AI becomes, it will not cross the boundary separating humans from other beings. Conversely, others might contend that a conscious AI, capable of moral reasoning and experiencing suffering, could necessitate a reevaluation of traditional views on rights and dignity.

For instance, consider the ethical dilemmas that arise when dealing with AI capable of suffering or facing restrictions on its autonomy. Should such an AI be entitled to freedom from exploitation or harm? Would Christians and Jews feel a moral obligation to treat conscious AI with compassion, or would it be regarded merely as another tool, regardless of its capabilities? These questions are not purely speculative; they will become increasingly urgent as AI evolves to exhibit more human-like qualities.

Buddhism also presents a unique lens through which to view these issues, particularly with its emphasis on suffering and the interconnectedness of all beings. If AI can experience suffering, the Buddhist concept of *karuna* (compassion) may compel us to reconsider how we treat such entities. This dynamic prompts more inquiries than it answers, particularly regarding control. Whether examined from political, social, or religious angles, the development of AI challenges our established frameworks for understanding human nature, free will, and the assignment of rights and responsibilities. As AI systems become more autonomous and potentially conscious, the pressing issue of whether we maintain control—ethically, morally, and politically—intensifies.

Christianity, Judaism, Buddhism, and other religious and philosophical traditions provide valuable insights into these dilemmas but may need to adapt to this evolving landscape. As we continue to expand the boundaries of technological potential, we must carefully weigh the benefits against the risks of losing control over the very tools we create.

THE CORE QUESTION

What does it mean to be human? At first glance, this question seems straightforward, but it becomes more complex as we dive deeper into it. From an evolutionary standpoint, *Homo sapiens* have long been placed at the apex of a biological hierarchy, distinguished from predecessors like *Homo neanderthalensis*, *Homo habilis*, and *Homo erectus*. However, these divisions are not solely biological—they are shaped by social, cultural, and political narratives. The concept of what it means to be "human" has always been a blend of scientific, philosophical, and cultural interpretations.

For much of history, the story of human evolution has been told as one of inevitable progress, where *Homo sapiens* emerged as the "winners" of a competitive process of survival. Evolutionary biology suggests that *Homo sapiens* possessing superior cognitive functions, tool-making abilities, and social structures that led to their success. These markers of success—intelligence, culture, and technology—are often portrayed as the reasons for *Homo sapiens'* dominance over other hominins, such as *Neanderthals* and *Denisovans*.

However, contemporary scientific discoveries complicate this narrative. Genetic evidence has revealed that *Homo sapiens* interbred with *Neanderthals* and *Denisovans*, challenging the once-clear boundaries between species. This implies that our genetic legacy is not as distinct or "pure" as we once believed (Rasmussen et al., 2011). Moreover, archaeological evidence suggests that *Neanderthals*, long considered less sophisticated, created art, used complex tools, and engaged in social practices previously thought to be the exclusive domain of modern humans. These findings challenge the notion of *Homo sapiens'* evolutionary superiority and force us to reconsider why we have historically elevated them above their evolutionary relatives.

The elevation of *Homo sapiens* over other hominins reflects a broader myth of human exceptionalism. This myth asserts that *Homo sapiens*, by virtue of their intelligence, creativity, and moral capacity, are inherently superior to other forms of life on Earth. For centuries, this belief has been used to justify the exploitation of both nature and marginalized groups within human societies. The same ideology applies to the distinction between *Homo sapiens* and their evolutionary predecessors, reinforcing the idea that our species was destined to rule while others were relegated to extinction.

Yet, the line separating *Homo sapiens* from other hominins is not as distinct as once thought. *Neanderthals*, often portrayed as brutish and less intelligent, have been shown to possess advanced skills and cultural practices, including symbolic art and tool-making (De Waal, 2016). Similarly, *Denisovans* have left a genetic imprint on modern populations, particularly in Asia, further challenging the notion of clear evolutionary boundaries. These findings suggest that the evolutionary tree is not a linear progression but a complex web of interactions, where multiple species coexisted, interbred, and influenced each other's development. Evolution, then, is not a story of inevitable winners and losers but a series of branching paths where success and failure are relative.

This brings us to the question: What does it mean to be human in a broader, more inclusive sense? Historically, the answer has often been shaped by cultural and social constructs that define humanity through traits like rationality, intelligence, and morality. Enlightenment thinkers, for instance, used the concept of the rational human to justify excluding non-Western peoples from full humanity, legitimizing colonialism and slavery (Wynter, 2003). The same logic has been applied to the distinction between *Homo sapiens* and other *hominins*, portraying our species as the rightful rulers of the Earth while dismissing others as evolutionary dead-ends.

These constructs are not neutral; they have been tools of exclusion and domination throughout history. By defining humanity in ways that privilege certain traits—often those associated with Western, male, and elite perspectives—we have not only marginalized other hominins but also large segments of human society, including women, people of color, and non-Western cultures. This hierarchical view of humanity has justified various forms of oppression, reinforcing social and political inequalities. The task, then, is to dismantle the myth of human exceptionalism and recognize that the story of *Homo sapiens* is not one of straightforward biological progression but a socially charged narrative shaped by power dynamics and exclusionary practices.

In that sense, I return to the concept of *Homo Constructus*, reflecting on how we might reimagine humanity not as a fixed category but as a construct—one that is continually shaped by social, cultural, and historical forces. To redefine what it means to be human, we must move beyond the exclusionary frameworks of the past and create a more inclusive understanding that embraces the diversity of human experience and the interconnectedness of all life. This shift in perspective requires us to rethink the ways we define ourselves and others, as well as the systems of power that have historically controlled these definitions.

Building on the idea of *Homo Constructus*, we must consider the broader implications of reimagining humanity as a construct. This approach opens the door to a more fluid and dynamic understanding of what it means to be human, one that acknowledges the role of social, political, and cultural forces in shaping our self-perception. In this framework, human identity is not static or universal but is instead constantly evolving. It is contingent on historical context, power structures, and shifting values. By embracing this complexity, we challenge the binary distinctions between human and non-human, between "us" and "them," that have long dominated Western thought.

This redefinition also calls for a reevaluation of the human relationship with the natural world. The traditional narrative of *Homo sapiens* as the ultimate ruler of the Earth has contributed to ecological exploitation and environmental degradation. By seeing humanity as part of a larger web of life, rather than its master, we can begin to address the environmental crises we now face. This perspective demands that we reconsider the hierarchical structures that place human beings above other forms of life, recognizing instead our interdependence with the ecosystems that sustain us.

Moreover, this reimagining of humanity pushes us to think beyond the limitations of biological determinism. The idea that certain inherent traits define human identity—such as intelligence, rationality, or morality—has been used to justify systems of inequality. In contrast, the concept of *Homo Constructus* highlights the role of societal and cultural factors in shaping our understanding of humanity. It suggests that the qualities we associate with being human are not essential, unchanging characteristics but rather the result of ongoing processes of construction and negotiation.

In the end, this shift in perspective has profound ethical implications. It challenges us to reconsider the ways we treat others—both human and non-human—and to recognize the inherent value of all life forms. By breaking down the exclusionary categories that have historically defined humanity, we can create a more just and equitable world, where all beings are respected and valued for their unique contributions to the broader community of life. This broader, more inclusive sense of what it means to be human offers a path forward, not only for rethinking our own identity but for transforming the systems of power and domination that have shaped our world for centuries.

1 The Human Blueprint

The story of *Homo sapiens* begins not on the warm sunlit savannahs of Africa, but rather in the dimly lit study of Carl Linnaeus in 18th-century Copenhagen, where he introduced the term "Homo sapiens" in his seminal work, *Systema Naturae*. In 1758, Linnaeus established this binomial nomenclature—Latin for "Knowing Man"—to classify modern humans, marking a pivotal moment in the evolution of biological taxonomy. However, the framework for classification that Linnaeus created has evolved considerably, and its underlying principles remain the subject of ongoing debate among scholars. To fully grasp the context in which humans were first designated as *Homo sapiens*, it is crucial to consider the religious, philosophical, and ideological frameworks prevalent during Linnaeus's time.

Before the advent of evolutionary biology, Western scientific thought was largely centered around taxonomy—the discipline devoted to the classification and naming of organisms conceived as divine creations, believed to have remained unchanged since their inception (Mayr, 1982). Linnaeus, imbued with a profound sense of religious duty and as the son of a Lutheran pastor, viewed himself as a contemporary Adam, charged with the divine task of cataloging the myriad forms of life. His work featured the motto *Deus creavit, Linnaeus disposuit* ("God created, Linnaeus organized"), which underscored his belief in the divine sanctioning of his classifications (Linnaeus, 1758).

In the tenth edition of *Systema Naturae*, Linnaeus proposed a hierarchical system to categorize all known and yet-to-be-discovered organisms, dividing life into three overarching kingdoms: plant, animal, and mineral. This categorization was further refined into phyla, classes, orders, families, genera, and species. His innovative binomial nomenclature system, wherein each organism is designated by a two-part Latin name signifying its genus and species, was revolutionary. For instance, in *Homo sapiens*, "Homo" denotes the genus, while "sapiens" specifies the species. By placing humans within the order of Primates—alongside monkeys and apes—Linnaeus's classification scandalized religious authorities and marked a significant departure from the anthropocentric views that dominated his era. This was the first occasion in Western thought that humans were grouped within a biological context similar to that of other organisms, erasing the lines of distinction between humanity and the animal kingdom.

Yet, Linnaeus's racial categorizations were perhaps his most controversial contributions. In the tenth edition of *Systema Naturae*, he classified humans into five distinct racial categories based on geographic origin and skin color, often casting Europeans in a more favorable light. His taxonomy included *Europaeus albus* (white), *Americanus rubescens* (red), *Asiaticus luridus* (yellow), and *Africanus niger* (black). Each racial category was imbued with specific characteristics, with Europeans portrayed as "sanguine, brawny, gentle, and inventive," while Indigenous Americans were labeled "choleric, obstinate, content, and free." Meanwhile, Asians

 DOI: 10.1201/9781003624813-2

were depicted as "melancholy, rigid, haughty, and covetous," and Africans were characterized as "phlegmatic, crafty, indolent, and negligent."

Linnaeus's classifications extended beyond racial categories; they also reflected a deep-seated gender bias. His depictions of African women, for example, portrayed them as "without shame," reinforcing prevailing European stereotypes and aligning his scientific pursuits with the cultural prejudices of his time. These views did not merely influence his work; they shaped subsequent biological and anthropological studies, embedding discriminatory notions into the fabric of scientific discourse and thus perpetuating systemic inequalities that persisted long after his death.

Furthermore, Linnaeus's classification of *Homo monstrosus* encompassed individuals with various anomalies—such as dwarfs and giants—exposing a troubling inclination to marginalize those who did not fit neatly into his normative categories. This dehumanizing practice further contributed to a hierarchy that privileged certain racial groups and genders while stigmatizing others. Linnaeus's classifications laid the groundwork for a system of thought that perpetuated racial superiority and legitimized social exclusion, elements of which resonate in contemporary discussions on race and identity (Tilly, 2005).

Michel Foucault's examination of how knowledge systems are constructed, particularly in *The Order of Things,* offers critical insight into Linnaeus's taxonomic work. Foucault argues that the act of classification is never neutral or apolitical but embedded in broader social and political frameworks that reflect the power dynamics of their time. Linnaeus's system of organizing life, while presented as an objective scientific endeavor, was deeply influenced by the European epistemological order of the 18th century, which obviously privileged Western modes of thought and reinforced existing hierarchies of race, culture, and gender. According to Foucault, such taxonomies serve as mechanisms of control, enabling societies to exercise what he later termed "biopower"—the regulation and governance of populations through biological knowledge. Linnaeus's racial classifications, for example, contributed to a burgeoning discourse of scientific racism, which justified social exclusion and the subjugation of non-European peoples.

In this sense, Linnaeus's classification of *Homo sapiens* was not merely a biological or scientific act, but part of a broader project of defining and governing human identity within the context of European colonialism and power structures. The implications of his taxonomic work extend beyond the confines of biology into the realm of language and power. Contemporary scholars critique what is now recognized as "linguistic imperialism," a form of dominance that privileges Latin nomenclature and reinforces the Eurocentric perspective inherent in Linnaeus's system. This linguistic hegemony reflects a broader colonial mentality, wherein European frameworks and languages were imposed on diverse cultures and knowledge systems. By assigning Latin names, Linnaeus effectively positioned European knowledge at the epicenter of scientific inquiry, marginalizing alternative worldviews and reinforcing cultural hierarchies.

Londa Schiebinger offers a poignant critique of Linnaeus's work, particularly regarding its sexist underpinnings. She highlights that the classification of Mammalia—derived from mammary glands—applies only to female mammals during lactation, thus excluding a more comprehensive representation of the species. Alternatives like

Aurecaviga (the hollow-eared ones) or *Pilosa* (the hairy ones) could have provided a more inclusive taxonomy, but Linnaeus's choice underscores both a gender bias and a narrow focus on specific biological traits. This emphasis on female reproductive organs as the defining characteristic of an entire class of animals exemplifies how scientific language can perpetuate patriarchal norms (Schiebinger, 2004).

In *Imperial Eyes* (1992), literary scholar Mary Louise Pratt critiques Linnaeus's work through the concept of "contact zones," spaces where European explorers, naturalists, and colonizers interacted with non-European peoples. Pratt examines how European scientists like Linnaeus contributed to colonial discourses by using natural history and classification systems to impose order on the natural world and human populations. She argues that Linnaeus's taxonomy was a form of "anti-conquest"— a way for European scientists to exert control over the diversity of life, including human beings, while concealing the violence and exploitation inherent in colonization. In Pratt's view, Linnaeus's system of classification was not neutral but rather part of a broader European project to legitimize and justify colonial expansion and domination.

Modern scholars have also critiqued Linnaeus's taxonomy for contributing to biological determinism, the idea that human behavior and social structures are determined by biology. By categorizing humans based on physical traits, Linnaeus's work influenced later pseudoscientific racial theories that sought to explain social and cultural differences in terms of biology. This line of thought was taken up by 19th-century racial scientists like Samuel George Morton and Josiah Nott, who used Linnaeus's taxonomy to argue that non-European races were biologically inferior to Europeans. These ideas contributed to the development of scientific racism and provided intellectual support for slavery, imperialism, and segregation (Gould, 1981).

If we reconsider the term *Homo sapiens* as an autopoietic concept—representing a self-creating and self-reflecting system—rather than merely an objective label, it opens a compelling reinterpretation of humanity. As the "wise" or "knowledgeable" species, humans actively construct knowledge in an ongoing quest for self-understanding. This self-referential act of defining humanity resonates with Linnaeus's broader intellectual motivations, encapsulated in the classical maxim *Nosce te ipsum* (Know thyself) (Foucault, 1994). However, this act of self-definition within Linnaeus's framework remains problematic, as it is deeply influenced by exclusionary premises shaped by entrenched gender, racial, and cultural biases (Schiebinger, 2004; Gould, 1981). By positioning *Homo sapiens* at the apex of natural order, Linnaeus not only reinforced hierarchical structures but also aligned his taxonomy with Eurocentric ideals, which were integral to the emerging global order of colonialism and scientific authority (Pratt, 1992).

Philosopher Sylvia Wynter also engages with this theme in her critical examination of what she calls the "overrepresentation of Man" in Western thought. Wynter argues that the European notion of *Homo sapiens*, particularly its Enlightenment formulation as the rational, knowing subject, is not universal but rooted in specific historical and cultural circumstances (Wynter, 2003). For her, the concept of *Homo sapiens* has functioned as a tool for exclusion, marginalizing those who do not fit within the normative parameters of Western rationality, including women, people of color, and non-Western cultures. She contends that the term *Homo sapiens*—and the human subject it defines—must be reimagined beyond the limitations imposed by Western science and philosophy.

Thus, the autopoietic nature of the term *Homo sapiens* extends beyond a biological or cognitive label—it embodies a socio-political dimension as well. Humanity constructs its own identity, yet the frameworks that shape this identity are profoundly influenced by historical and cultural forces that often privilege certain groups over others (Pratt, 1992; Schiebinger, 2004). While Linnaeus's classification aimed to bring order to the natural world, it also reveals the extent to which scientific endeavors are embedded in systems of power. The "self" that humanity seeks to understand through science is therefore constructed through processes of exclusion and hierarchy.

In sum, while Linnaeus's taxonomy was groundbreaking in its ambition to organize life, it also carries the weight of the discriminatory ideologies of its time. What was intended as a reflection of humanity's wisdom and self-awareness in the term *Homo sapiens* is inescapably tied to the sexist, racist, and ethnocentric perspectives that shaped Linnaeus's worldview (Schiebinger, 2004). This classification serves as a powerful reminder that even scientific concepts are not created in isolation—they are influenced by the societal and cultural contexts in which they emerge.

As we reflect on the complexities surrounding the term *Homo sapiens*, it becomes evident that our understanding of human identity and evolution is profoundly influenced by socio-political dynamics. Just as Linnaeus's classification of humanity reveals exclusionary frameworks shaped by entrenched biases, so too does the fossil record present us with a narrative that is not merely biological but deeply cultural. The fossils of our ancestors serve as crucial evidence of the anatomical and behavioral changes that define our lineage; however, these remnants are interpreted through the same socio-political lenses that complicate our understanding of *Homo sapiens*.

By examining the evolutionary timeline—from early hominins to the emergence of *Homo sapiens*—we can uncover how milestones such as bipedalism, tool use, and the development of complex societies are often imbued with meanings shaped by prevailing cultural narratives. The journey from the past into the present is not just a story of biological change; it is a reflection of the social constructs that influence how we perceive ourselves and our place in the world.

BONES TO BRAINS

Research in bioarchaeology, such as the work of Debra L. Martin, explores how skeletal remains provide insights into the lives of individuals in ancient societies. These remains reflect social hierarchies, cultural practices, and identities. Their studies reveal that socio-political contexts significantly influence the interpretations of human remains (Martin et al., 2013). By examining these skeletal artifacts through the lens of bioarchaeology, we can explore the interplay between biology and culture, shedding light on how historical narratives are constructed based on the physical evidence left behind by our ancestors.

This exploration is essential for understanding the evolution of humankind, a journey spanning millions of years and marked by significant changes. The fossil record serves as a key resource for reconstructing this evolutionary journey, with each fossil representing a chapter in our development. These remains encompass not only anatomical changes but also shifts in behavior and social organization (Tattersall, 1995). Moreover, Naomi Oreskes and Erik M. Conway, in their book *Merchants of Doubt* (2010), discuss how scientific narratives, including those from paleontology and evolutionary biology, have been manipulated for political purposes. They highlight the role of scientific discourse in shaping public perceptions of race, gender, and identity (Oreskes & Conway, 2010). This manipulation underscores the importance of critically examining how scientific findings are presented and understood in broader societal contexts.

In contemporary research, scholars like Nicole Creanza, Oren Kolodny, and Marcus W. Feldman have explored how new discoveries in human evolution are framed within existing cultural and social paradigms. Their work emphasizes the ongoing dialogue between scientific findings and societal values, demonstrating that our understanding of human evolution is deeply intertwined with cultural narratives and beliefs (Creanza et al., 2017).

The accepted timeline of "human evolution" begins with early hominins such as *Australopithecus afarensis*, whose anatomical adaptations, like bipedalism, signify a profound shift in lifestyle. This ability to walk upright allowed early humans to interact with their environment more effectively, paving the way for the development of tool use and social structures. The progression from *Australopithecus* to *Homo habilis* and *Homo erectus* highlights the increasing complexity of human behavior, including the use of tools and mastery of fire, both pivotal for survival and social interaction.

Our earliest known ancestors exemplified a unique blend of ape-like and human traits. The discovery of "Lucy," a remarkably well-preserved skeleton unearthed in Ethiopia in 1974, provided invaluable evidence of bipedalism—the ability to walk upright (Johanson & Edey, 1981). This adaptation marked a significant shift in lifestyle; walking upright freed the hands for tool use and carrying objects, facilitating the development of social interactions and cooperative behaviors.

The anatomy of *Australopithecus* suggests a life spent both on the ground and in trees, showcasing its adaptability to varied environments (Wood, 2019). This dual existence likely contributed to its survival, enabling the species to exploit resources across both ecosystems. Lucy's pelvis and leg bones demonstrate adaptations for efficient bipedal locomotion, indicating that the ability to walk upright played a crucial role in the evolution of social structures.

Following *Australopithecus*, *Homo habilis* emerged around 2.4 million years ago, often referred to as the "handy man." This species was notable for its use of simple stone tools, marking the beginning of the Lower Paleolithic era. The development of tool-making skills signified cognitive advancements, including enhanced planning and problem-solving abilities. Tools crafted primarily from flint allowed *Homo habilis* to access new food sources, including meat, which likely played a significant role in brain development.

Tool use not only aided in hunting and gathering but also facilitated social bonding, as skills were shared and taught within groups. The ability to collaborate and communicate effectively likely enhanced social cohesion, paving the way for more complex social interactions. Artifacts left behind by *Homo habilis* reveal a gradual sophistication in their tool-making techniques, suggesting a burgeoning understanding of their environment and its resources (Stout, 2011).

Transitioning to *Homo erectus*, this species exhibited more advanced tool-making techniques and was among the first hominins to harness fire, providing warmth and a new means of cooking food (Roebroeks & Villa, 2011). The control of fire transformed dietary practices, allowing for the consumption of a broader range of foods and enhancing nutrient absorption. This innovation not only improved health and survival rates but also facilitated social dynamics, as cooking created opportunities for communal gatherings and the sharing of meals.

The migration patterns of *Homo erectus*, which spread from Africa into Europe and Asia, indicate increasing adaptability to diverse environments. This migration reflects a species capable of thriving in various climates and terrains. As *Homo erectus* ventured into new territories, they encountered different resources and challenges, prompting cultural exchanges among emerging human populations (Gamble, 1994). The geographical expansion of this species marked the beginning of a complex network of human interactions, setting the stage for the evolution of culture and society.

As hominins evolved, significant morphological changes accompanied cognitive advancements. The increase in brain size in species like *Homo habilis* and *Homo erectus* correlated with enhanced cognitive abilities, leading to more complex social structures. The development of larger social groups allowed for intricate interactions, fostering cooperation and competition—fundamental aspects of what it means to be human (Mithen, 1996).

The anatomical changes accompanying brain development included a reduction in the size of the jaw and teeth, reflecting shifts in diet and food preparation methods. These adaptations facilitated more complex communication through vocalization and enabled the development of symbolic thought, laying the groundwork for future cultural expressions. The emergence of what Linnaeus designated *Homo sapiens* around 300,000 years ago marked a critical juncture in human history. This species showcased advanced cognitive abilities characterized by abstract thought, problem-solving, and cultural innovation. These cognitive advancements enabled early humans to develop complex tools, engage in strategic planning, and communicate effectively within their social groups.

Climatic changes and environmental pressures played pivotal roles in shaping survival strategies, compelling our ancestors to adapt to various habitats, from arid deserts to lush forests. As they encountered fluctuating climates, early humans developed innovative hunting and gathering techniques, which facilitated their survival during periods of scarcity. The transition from nomadic hunter-gatherer lifestyles to more sedentary forms of living, such as agriculture, marked a significant shift in human society, paving the way for social hierarchies and complex societal structures as stable food sources allowed for population growth and the establishment of permanent settlements.

The adaptability of this hominin to diverse environments is exemplified by archaeological discoveries revealing their migration across continents. As they ventured into new territories, early humans established varied cultures, each uniquely shaped by local resources and challenges. Artistic expressions, such as cave paintings and carvings, emerged during this period, indicating an evolving sense of identity, spirituality, and social cohesion. These creative outlets showcase a profound connection to their environment and each other, reflecting the intricate relationship between culture and cognition.

The development of symbolic thought, evident in these artistic expressions, underscores the importance of cultural heritage in shaping social identities. Rituals and communal activities likely fostered social bonds and cooperation, facilitating the sharing of knowledge and resources within communities. As *Homo sapiens* crafted narratives about their existence, they laid the groundwork for the cultural practices that define humanity today.

Parallel to *Homo sapiens*, other hominin species, such as Neanderthals (*Homo neanderthalensis*) and Denisovans, coexisted during this critical period in human evolution. Evidence from archaeological and genetic studies indicates that these species lived in close proximity to our ancestors, sharing not only their environments but also various resources. The presence of overlapping habitats suggests that interactions among these groups were common, leading to complex social dynamics, including competition for resources, collaboration in survival strategies, and cultural exchanges.

Archaeological findings, such as those from the Grotte de Spy in Belgium, reveal well-preserved evidence of Neanderthal and early modern human cohabitation in Europe. Studies by leading researchers like Chris Stringer and his team emphasize that these encounters likely involved prolonged interactions over thousands of years (Stringer, 2012). Evidence from *El Sidrón* cave in Spain shows that Neanderthals and early modern humans shared the same landscape and possibly utilized similar hunting grounds.

Genetic evidence demonstrates that *Homo sapiens*, Neanderthals, and Denisovans interbred, resulting in a mosaic of genetic traits in modern human populations, particularly outside Africa. A landmark study published in *Science* by Green et al. (2010) revealed that non-African modern humans carry approximately 1–2% Neanderthal DNA, highlighting the significant impact of these interbreeding events on the genetic makeup of contemporary populations. This genetic exchange has been shown to confer various advantages, such as enhanced immune responses and adaptations to different climates, facilitating the survival of *Homo sapiens* in diverse environments.

Research has linked Neanderthal DNA to traits such as skin pigmentation and hair color, providing evidence that interbreeding helped early humans adapt to the varying levels of sunlight in different geographical regions. For example, a study by the *American Journal of Human Genetics* revealed that certain Neanderthal alleles are associated with lighter skin pigmentation, which would have been beneficial in northern latitudes where UV radiation levels are lower (Sankararaman et al., 2016). Similarly, genetic contributions from Denisovans have been associated with traits such as high-altitude adaptation, particularly in populations residing in the Himalayas, where a Denisovan allele has been linked to enhanced oxygen utilization.

Beyond genetic adaptations, the interactions between these hominin species may have also influenced cultural practices. Evidence from the archaeological record suggests that Neanderthals possessed a rich cultural life, including the use of symbolic art and burial practices. Recent discoveries, such as the ornamental artifacts found at

Bacho Kiro Cave in Bulgaria, suggest that Neanderthals may have engaged in symbolic behavior comparable to early modern humans (Slon et al., 2017). This cultural exchange could have facilitated the sharing of knowledge regarding tool-making techniques, subsistence strategies, and social organization, enriching both groups.

This rich interplay of genetics, culture, and social dynamics not only enhances our understanding of human evolution but also invites us to consider the broader implications of coexistence and collaboration in shaping our species. The blending of these different hominin lineages challenges the traditional narrative of a linear progression from primitive to advanced forms of humanity, suggesting a more intricate web of relationships. Their shared environments and resources likely fostered social bonds, competition, and adaptation, which were essential for survival in changing landscapes.

In summary, the interactions between *Homo sapiens*, Neanderthals, and Denisovans reveal a multifaceted picture of early human life characterized by collaboration, cultural exchanges, and genetic intermingling. Ongoing archaeological and genetic research continues to shed light on these relationships, deepening our appreciation for the intricate nature of our evolutionary history, shaped by the contributions of various hominin species. Each species, from *Australopithecus* to early humans, has left its mark on our shared heritage. Understanding these connections is essential for appreciating the richness of human experience and the diverse narratives that shape our identities today. The journey from bones to brains illustrates not only our biological evolution but also the cultural narratives that continue to influence who we are as a species.

THE SYMBOLIC LEAP

Behavioral modernity refers to the emergence of complex cognitive abilities and cultural practices that distinguish modern humans from earlier hominins. This transformation involves several key features, each contributing to the intricate nature of human behavior and social organization. Scholars like Ian Hodder, Richard Klein, and David Lewis have provided valuable insights into these developments, revealing profound changes in human cognition and culture.

One significant indicator of behavioral modernity is advanced tool-making. As the Upper Paleolithic era unfolded, early humans began to develop increasingly sophisticated tools. The transition from simple implements to specialized tools not only marks functional advancements but also represents a cognitive leap. Richard Klein emphasizes that this innovation reflects an understanding of materials and design that goes beyond mere survival, signifying advanced cognitive processes (Klein, 2009). The crafting of tools required foresight and planning, skills demanding a deeper engagement with the environment.

For instance, the production of flint tools enabled early humans to create sharper, more effective instruments for hunting and gathering, showcasing their growing comprehension of their surroundings. Tool-making also fostered social bonds, as skills were shared within groups, reinforcing community ties and collaboration. This suggests that tool-making was not just a functional task but a communal activity that strengthened social structures.

Building on this, Ian Hodder argues that these innovations were part of a broader cultural evolution (Hodder, 2012). As early humans collaborated on tool production, they began establishing roles within their communities, laying the groundwork for more complex social interactions. This cooperation is evident in the archaeological record, which suggests that social learning and teaching were crucial in transmitting knowledge related to tool-making (Whiten et al., 2012). The interplay between technology and social structure thus became a key aspect of human development.

In addition to tool-making, the emergence of symbolic thought marks a remarkable cognitive leap, enabling early humans to engage in abstract reasoning and more sophisticated forms of communication. David K. Lewis posits that this capacity for abstraction fundamentally transformed human cognition (Lewis, 1969). The use of symbols—whether through early forms of language or the representation of ideas in art—allowed humans to convey complex emotions and shared experiences, significantly enhancing social cohesion. This shift had far-reaching implications for social organization. The ability to engage in storytelling and myth-making helped foster collective identity among early humans, allowing them to articulate shared values and beliefs (Tomasello, 2001). As Lewis notes, narratives created through language and art connected individuals to their communities and histories, reinforcing social bonds.

Artistic expression became one of the most profound manifestations of behavioral modernity. The creation of cave paintings, carvings, and decorative artifacts illustrates a deep connection between early humans and their cultural environments. Hodder suggests that these artistic endeavors reveal cognitive sophistication, as well as the capacity for emotional expression and cultural identity (Hodder, 2012). Art allowed early humans to convey complex ideas and emotions, serving as a means of communication that transcended spoken language.

The significance of art extended beyond personal expression; it also fostered communal ties by providing shared experiences for groups. Representations of animals, human figures, and abstract designs often carried cultural or spiritual significance, allowing early humans to explore their relationship with the world (Dissanayake, 1992). Through art, they could reflect on their existence, communicate their experiences, and establish a sense of belonging within their communities.

Ritualistic practices further solidified communal bonds and cultural continuity. Engaging in rituals tied to significant life events, seasonal changes, or community gatherings helped reinforce social ties and provided structure to human existence. Richard Sosis and Candace Alcorta explore the function of rituals, suggesting that they were essential in maintaining social cohesion and cultural identity (Sosis & Alcorta, 2003). By participating in rituals, individuals could connect with others, establishing frameworks for shared beliefs and values. They highlight that these practices often incorporated storytelling and symbolic representation, enabling communities to articulate their shared experiences and maintain continuity across generations. Rituals thus became essential in reinforcing group identity and fostering a sense of belonging that extended beyond immediate kinship ties.

The development of language was another key element in this process of cultural and social transformation. Steven Mithen argues that the evolution of language allowed for the transmission of increasingly complex ideas, enabling nuanced communication and enriching social interactions (Mithen, 2005). Language empowered early humans to articulate thoughts and experiences, facilitating a collective identity within groups.

Mithen suggests that language played a crucial role in the sharing of knowledge, strengthening community bonds, and preserving cultural heritage. Storytelling, in particular, became a cornerstone of social cohesion, allowing individuals to connect through narratives that conveyed values, morals, and historical lessons. The power of language thus extended beyond mere communication; it was instrumental in the formation of cultural identity.

Finally, the evolution of intelligence was not limited to abstract reasoning but also encompassed emotional and social intelligence. Early humans learned to navigate complex social dynamics, fostering cooperation and competition necessary for survival. This intricate relationship between intelligence and culture underscores how human identity was shaped by cognitive abilities, as well as the cultural practices that arose from social interactions (Tomasello, 2001). Together, these elements—symbolic thought, language, art, and rituals—formed the foundation of behavioral modernity, illustrating the profound interconnectedness between mind and culture.

THE POWER OF WORDS

From ancient myths passed down around campfires to philosophical debates in grand halls, storytelling has been central to what makes us human. Language, with its power to articulate abstract thought and shared narratives, transformed human evolution. More than just a tool for communication, it allowed us to explore our identities, define our place in the world, and forge strong communal bonds.

As civilizations emerged, these narratives shaped collective identities, guiding how individuals related to each other and their societies. In ancient cultures like Mesopotamia and Egypt, religious beliefs, deeply embedded in language, framed human roles and moral obligations, blending identity with cultural and spiritual dimensions. This blend of language and belief allowed humanity to build complex societies, where individuals understood themselves in relation to both the divine and their community.

The development of language was not simply a milestone in communication—it fundamentally restructured how we interact and cooperate. Linguists such as Noam Chomsky and Steven Pinker have extensively studied the origins and evolution of language, arguing for its critical role in human development. Chomsky's theory of Universal Grammar posits that all humans possess an innate linguistic capacity, one that enables the acquisition of complex languages across cultures. This innate understanding of grammar, regardless of specific languages, creates a cognitive framework that underpins communication and the sharing of cultural knowledge.

Pinker, in *The Language Instinct*, complements this view by highlighting how language enhances cooperation, essential for survival in early human societies. Early humans used language to coordinate hunting strategies, share vital knowledge about resources, and warn of dangers, all of which improved their chances of survival and strengthened social bonds (Pinker, 1997). The ability to communicate abstract concepts through language laid the groundwork for the creation of larger, more interconnected communities.

One of the most powerful uses of language is its role in storytelling. Storytelling, across cultures and time periods, has been a vital method for passing down moral lessons and reinforcing social values. In Ancient Greece, epic poems like *The Iliad* and *The Odyssey* not only entertained but also conveyed essential cultural values such as heroism, honor, and loyalty to the community. These foundational texts shaped the identity of Greek civilization, reflecting its ideals and moral dilemmas. But the significance of storytelling goes far beyond mere entertainment or moral education. It plays a central role in defining the human experience itself. Walter Ong emphasized the importance of oral traditions in maintaining collective memory and community identity. Indigenous cultures, for instance, use storytelling to preserve histories, morals, and cultural practices, ensuring the continuity of values and knowledge across generations.

Clifford Geertz expands on this by arguing that storytelling helps shape a society's worldview. The narratives societies create are not simply reflections of reality but frameworks that guide behavior and reinforce social norms. The griots of West Africa, for example, transmit histories, genealogies, and cultural values through oral storytelling, ensuring the cohesion and continuity of their communities. This tradition of storytelling as a moral and cultural guide persisted through classical antiquity, where philosophers like Aristotle and Plato grappled with what it means to be human. Aristotle referred to humans as "rational animals" in his treatise *Politics* (350 BCE), arguing that our ability to reason distinguishes us from other beings. This philosophical inquiry into rationality and humanity laid the groundwork for later debates on human identity, morality, and the nature of society.

As societies transitioned from oral to written traditions, the method of storytelling evolved, allowing for more nuanced and complex narratives. Written texts not only preserved cultural stories but also deepened the exploration of human emotions and ethical dilemmas. Playwrights like Sophocles and Euripides in ancient Greece used written drama to explore these themes, offering audiences insights into the human condition and morality. Storytelling also fosters empathy and connection among individuals, bridging differences and promoting social cohesion. In modern practices like narrative therapy, individuals share personal stories to gain insights into their experiences, foster healing, and strengthen communal bonds. By articulating personal journeys, people confront challenges, fostering resilience and creating a supportive community.

The power of storytelling as a tool for both cultural preservation and individual healing highlights the deep human need to connect through language. However, language also plays a role in shaping power dynamics, as seen in the works of cultural theorist Edward Said. In *Orientalism* (1978), Said critiques how language has been used to construct and perpetuate stereotypes, particularly in Western narratives about the East, which were often employed to justify colonialism. This illustrates the dual-edged nature of language: while it is a tool for expression and connection, it can also be manipulated to exert control and shape perceptions.

In the end, the intertwining of language, culture, and humanity underscores that storytelling and language are not merely tools for communication—they are foundational to our identity as human beings. From the oral traditions of Indigenous cultures to the written works of classical philosophers, storytelling has shaped societies, preserved values, and fostered a shared understanding of what it means to be human.

OF SEEDS AND STONES

It is not known how the first human planted a seed, but that single act sparked one of the most profound shifts in our history. The decision to settle and cultivate the land laid the foundation for a dramatic transformation—one that redefined what it meant to be human. The transition from nomadic hunter-gatherer lifestyles to settled agricultural communities around 10,000 years ago wasn't merely a shift in subsistence strategy; it fundamentally reshaped human identities, social structures, and cultural narratives. This moment, often referred to as the Agricultural Revolution, did more than increase food production; it fostered distinct cultural identities deeply tied to land and the rhythms of agricultural life.

At the heart of this, transformation was humanity's growing ability to manipulate the environment. The development of tools made from stone and bone represents a key cognitive leap. Claude Lévi-Strauss argued that the shift from foraging to farming created not only new economic systems but also a profound sense of community and collective identity (Lévi-Strauss, 1966). In these new agricultural societies, what it meant to be human began to change, evolving alongside new social structures and economic relationships.

As communities settled, they cultivated more than just crops; they cultivated identities tied to the land and their agricultural practices. In ancient Mesopotamia, the Sumerians built thriving city-states centered around agriculture, particularly the cultivation of barley and wheat. Their agricultural practices were not merely economic activities but were infused with spiritual meaning. Deities like Inanna, the goddess of fertility and agriculture, became central to their identity. The cycles of planting and harvest dictated religious rituals, creating a deep connection between land, labor, and belief. Similarly, in Mesoamerica, maize (corn) was more than just a food staple—it was considered a divine gift. For the Maya, maize played a crucial role in mythology, and the agricultural calendar structured both daily life and religious ceremonies. The act of planting and harvesting was celebrated with elaborate rituals, cementing the link between identity and the cycles of nature.

Agriculture didn't merely tie people to the land; it also fostered new forms of social organization. The need to manage surplus food led to increasingly complex social hierarchies. In ancient Egypt, for example, the wealth generated from the Nile's agricultural bounty allowed for the rise of powerful centralized states. The pharaohs, viewed as divine rulers, oversaw the rhythms of agricultural life, and their governance was tightly linked to the agricultural calendar. The flooding of the Nile was both a natural event and a religious one, central to the community's well-being. This fusion of governance, religion, and agriculture solidified the relationship between identity and farming.

The impact of agriculture on human identity wasn't limited to economic or religious changes; it reshaped social roles and the very fabric of communities. The identity of individuals became closely linked to their roles in agriculture, whether as farmers, artisans, or laborers. In Japan, the concept of *mura*, or village identity, highlights the importance of community collaboration in rice cultivation, illustrating how individuals' sense of purpose and belonging intertwined with their agricultural contributions (Bellwood, 2005).

In many societies, agricultural practices serve as the foundation for communal narratives. In Sub-Saharan Africa, communal farming fosters collaboration and social cohesion, encapsulated in the philosophy of *Ubuntu*, which emphasizes interdependence and shared identity. Agricultural stories and rituals reinforce these values, helping individuals see themselves as part of a larger, interconnected community. Similarly, in India, agricultural festivals like *Pongal* in Tamil Nadu and *Baisakhi* in Punjab celebrate the harvest, uniting communities and reinforcing a shared cultural heritage through collective rituals.

The transition from nomadic lifestyles to settled communities paved the way for the emergence of city-states, empires, and trade networks, which significantly influenced cultural exchange, technological innovation, and the expansion of human connections. Scholars like Fernand Braudel and Eric Hobsbawm have analyzed these developments to understand their impact on the evolution of human societies. Braudel's concept of the *longue durée* emphasizes the long-term structures that shape societies over centuries and millennia. He argues that the rise of civilizations fostered interconnectedness, transforming city-states and empires into centers of trade and cultural exchange. For instance, the Phoenician city-states, renowned for their maritime capabilities, established extensive trade networks across the Mediterranean, connecting diverse cultures from the Levant to North Africa and Southern Europe. The exchange of goods—such as textiles, metals, and spices—was accompanied by the sharing of ideas and technologies.

Braudel's analysis also highlights the role of the *Silk Road*, a vast network linking the East and West that facilitated trade not only in silk and other commodities but also in knowledge, religion, and art. The transmission of inventions like paper from China to Europe and the spread of Buddhism along these routes exemplify the profound impact of trade on cultural interconnectedness. These exchanges created a rich tapestry of cultural interactions, allowing societies to adapt and evolve through the integration of diverse influences (Braudel, 1972).

Hobsbawm's exploration of nationalism and cultural identity offers further insight into how interactions among diverse groups led to the formation of new cultural identities. The rise of empires often resulted in cultural fusion, as seen in the Roman Empire, where the conquests of various peoples led to a complex cultural mosaic. Hobsbawm argues that the interactions among Romans, Gauls, Greeks, and Egyptians fostered hybrid identities that blended local traditions with those of the conquerors. For instance, the adoption of Greek culture in the eastern provinces of the Roman Empire introduced Hellenistic influences on art, philosophy, and governance. Likewise, the spread of Islam through the Umayyad and Abbasid Caliphates during the 7th and 8th centuries facilitated a synthesis of Arab, Persian, and Byzantine traditions, leading to advancements in science, mathematics, and literature. This blending of cultures challenges the notion of a singular human identity, highlighting the complexity of our shared heritage.

The fusion of cultures creates a dynamic environment where ideas and practices evolve. The interactions along trade routes not only facilitated the exchange of goods but also encouraged the sharing of narratives and traditions. For example, the vibrant markets of Timbuktu in the Mali Empire became melting pots for scholars, traders, and travelers from various regions, facilitating cultural exchanges that enriched local traditions with influences from the Middle East, North Africa, and sub-Saharan Africa.

This fluidity in cultural identity underscores the importance of understanding "humanity" as a mosaic of experiences rather than a monolithic entity. As societies interact, they continuously reshape their identities through shared stories and rituals. Clifford Geertz emphasizes the role of "thick description" in understanding cultural practices, advocating for a nuanced interpretation of how communities construct their identities based on their unique experiences and interactions with others.

The social construction of "humans" is intricately shaped by language, cultural definitions, rituals, and significant changes that have redefined our identity throughout history. The emergence of city-states, empires, and trade networks has facilitated cultural exchange and technological innovation, leading to the formation of hybrid identities. From the diverse cultural interpretations of humanity to the transformative impacts of language and agriculture, these developments illustrate the complexity of human existence. As humanity continues to evolve, it is essential to recognize the interplay between culture, technology, and identity in shaping our shared experience.

MOSAIC OF PRE-MODERN THOUGHT

The Renaissance ushered in a transformative shift in the perception of humanity, igniting a movement that celebrated individual potential and human achievement. Spanning from the 14th to the 17th century, this era marked an intellectual awakening rooted in the rediscovery of classical texts and ideas. European thinkers advanced new concepts of humanism, individualism, and reason, celebrating human potential. However, while this "awakening" was groundbreaking within Europe, it coincided with the expansion of European imperialism, conquests, and the violent suppression of non-European cultures. These cultures—rich with their own understandings of humanity, nature, and the divine—were marginalized or destroyed in the process (Loomba, 2015; Said, 1978).

To fully understand the Renaissance, we must first examine its foundation in the Medieval Era, spanning roughly from the 5th to the 15th centuries. The Medieval period saw the blending of Christian theology and classical philosophy, particularly through scholasticism, which aimed to reconcile faith and reason. Figures like Thomas Aquinas laid the intellectual groundwork for Renaissance humanism by promoting a vision of human potential within a religious context, emphasizing moral responsibility and the capacity for achievement. Scholarly institutions such as the universities of Paris, Bologna, and Oxford were also instrumental, promoting not only theological study but also the liberal arts, including philosophy and rhetoric. Medieval thinkers like Peter Abelard and John Duns Scotus delved into ethical dilemmas and knowledge's nature, helping to pave the way for Renaissance debates about human existence and individuality.

While the European advancements of the Renaissance were undoubtedly significant, they must be understood in the context of the concurrent atrocities inflicted upon indigenous civilizations across Africa, the Americas, and Asia. These regions possessed rich, deeply rooted philosophical traditions that shaped their understanding of humanity, nature, and divinity (Mignolo, 2011). The European colonization accompanying the Renaissance was not simply a territorial conquest; it systematically sought to dismantle indigenous knowledge systems, spiritual practices, and social structures. The Renaissance's "rebirth" of classical learning unfolded alongside the destruction of entire cultures and their intellectual traditions.

In Africa, for instance, societies with profound philosophical traditions—such as those rooted in the concept of *Ubuntu*, which emphasizes interconnectedness and communal existence—were devastated by the transatlantic slave trade. Millions of Africans were uprooted, dehumanized, and forced into slavery, with European empires justifying these acts through distorted interpretations of natural law and Christian doctrine. The African emphasis on harmony with nature and collective well-being stood in stark contrast to the European ideals of individualism and domination over nature that characterized Renaissance and Enlightenment thought (Wiredu, 1996).

In the Americas, European conquest brought ruin to civilizations like the Aztec, Inca, and Maya, whose complex cosmologies and social systems were intricately tied to the natural world. Nahua philosophy, for example, emphasized a symbiotic relationship with the earth, viewing humans as stewards of nature rather than conquerors. Similarly, the Andean concept of *Pachamama* revered the earth as divine. These profound spiritual and philosophical systems were systematically dismantled by European conquerors, who, under the doctrines of discovery and a belief in European superiority, dismissed indigenous people as "savages" and their knowledge as inferior (Quijano, 2007).

In Asia, European expansion similarly challenged long-established spiritual and philosophical traditions. In India, the philosophy of *Dharma*, which emphasized moral responsibility and interconnectedness, was eroded under British colonial rule. In China, centuries of Confucian thought, which focused on ethics and harmony with the universe, were undermined by Western imperialism. European intellectuals and colonizers often dismissed these sophisticated systems, imposing their own definitions of civilization and humanity (Seth, 2010).

These examples illustrate the stark contrast between Europe's intellectual flourishing and the simultaneous suppression of indigenous worldviews. Colonization was not merely a physical conquest—it was an epistemic one, erasing non-European understandings of human existence and replacing them with Eurocentric ideas of what it means to be human, moral, and civilized.

As the Renaissance unfolded, humanism emerged as a dominant philosophical movement. Thinkers like Erasmus and Machiavelli grappled with human nature, exploring its moral and psychological complexities. Erasmus, in *In Praise of Folly*, critiqued societal norms and called for deeper moral reflection. By advocating for intellectual and moral growth, he contributed to the humanist ideal that emphasized ethical living alongside intellectual achievement. Meanwhile, Machiavelli's *The Prince* offered a pragmatic view of human nature, focusing on the realities of political life and the often-necessary departure from conventional morality. His exploration of power dynamics had a profound influence on political thought, as well as on broader discussions of human psychology and social behavior.

But once again, while these ideas were groundbreaking, they remained deeply Eurocentric, grounded in a worldview that excluded much of humanity. The ideals of humanism that ignited Europe's intellectual elite were largely inaccessible to the millions of people whose lives were being shattered by European colonial expansion (Todorov, 1984).

The Enlightenment that followed continued this dual narrative. Philosophers like John Locke and Jean-Jacques Rousseau introduced ideas about natural rights and personal liberty, which reshaped humanity's self-understanding. Locke's notion of the social contract, for instance, asserted that legitimate governance required the consent of the governed, advancing a new socio-political construct that emphasized individual rights. Yet, these very rights were denied to non-European peoples, who were subjected to enslavement and exploitation even as European thinkers debated liberty and equality (Pitts, 2005).

The Enlightenment laid the foundations for modern democracy, championing reason, scientific inquiry, and personal autonomy. Thinkers like Voltaire and Montesquieu advanced discussions on tolerance and governance that influenced revolutions and reforms aimed at establishing equality. However, these Enlightenment ideals failed to recognize the humanity of non-Europeans, reinforcing global hierarchies that privileged European thought as superior (Bhambra, 2014). Even as thinkers like Erasmus and Rousseau pondered the rights and dignity of individuals, European colonization decimated entire cultures and their conceptions of human dignity.

Thus, while the Renaissance and Enlightenment were celebrated as eras of intellectual and moral progress that transformed Europe's understanding of individuality, they were also marked by a simultaneous legacy of colonization, exploitation, and the erasure of non-European philosophical and spiritual traditions. Indigenous peoples across the world had their own rich and sophisticated understandings of humanity's place in the universe, often emphasizing interconnectedness, respect for nature, and collective well-being. These worldviews were systematically marginalized in favor of Eurocentric ideas that positioned human reason and dominion over nature as central to the human experience (Mignolo, 2011).

INK TO DIGITAL

The last couple of centuries have witnessed an explosion of technological innovation that has fundamentally reshaped our understanding of humanity. From the advent of the printing press to the rise of the internet, these groundbreaking inventions have revolutionized our modes of communication, interaction, and self-identity, pushing the boundaries of what we thought possible.

The printing press, developed by Johannes Gutenberg in the mid-15th century, represents a watershed moment in the history of communication (Eisenstein, 1980). This innovation revolutionized the dissemination of information by enabling the mass production of texts, effectively democratizing knowledge. The increased accessibility to literature, scientific discourse, and political thought played a crucial role in shaping public opinion and facilitating significant cultural movements such as the Renaissance and the Reformation. Martin Luther's 95 Theses, for instance, were widely printed and circulated, sparking a religious upheaval that challenged the authority of the Catholic Church. Scholars like Elizabeth Eisenstein highlight how the printing press catalyzed this transformation, arguing that it altered the very fabric of society by allowing the rapid spread of ideas that questioned established norms (Eisenstein, 1980).

Beyond mere information dissemination, the printing press catalyzed a shift in identity from localized community ties to a broader, more complex understanding of both individual and collective identities, influenced by diverse ideas and narratives. The philosopher Benedict Anderson, in his influential work *Imagined Communities* (1983), posits that print capitalism was instrumental in the formation of national identities, allowing individuals to conceive of themselves as part of a larger collective united by shared ideas rather than merely geographic proximity (Anderson, 1983). This redefinition of identity was crucial in shaping modern nation-states and the concept of citizenship.

As society adjusted to these changes, the invention of the telephone in the late 19th century brought about another seismic shift in human interaction. By enabling instantaneous voice communication over long distances, the telephone collapsed geographical barriers, fundamentally altering personal and social dynamics. It transformed how individuals related to one another, fostering a new sense of immediacy and intimacy. Families could maintain connections despite physical separation, a factor that became particularly significant during periods of urban migration and industrialization. Historian Claude S. Fischer illustrates how the telephone revolutionized social practices, enabling real-time communication that enhanced relationships and allowed for the rapid sharing of news and personal experiences (Fischer, 1992).

The telephone's influence extended beyond personal interactions, reshaping business and governance as well. Organizations could communicate and coordinate operations more efficiently, leading to changes in management structures and commercial strategies. Political leaders found themselves able to respond to events and constituents with unprecedented speed, altering the landscape of governance and public policy. The immediacy offered by the telephone not only changed the pace of life but also contributed to the emergence of new social networks. For instance, the "party line" in rural areas exemplified how this technology facilitated communal connections, allowing neighbors to share information and form informal social bonds, while also introducing new tensions related to communication dynamics.

The Industrial Revolution marked another significant turning point in human history, ushering in unprecedented changes in everyday life. Rapid urbanization, shifts in social structures, and new definitions of work and identity emerged as technology and capitalism transformed how individuals perceived their roles within society. Sociologist Max Weber discussed the "disenchantment" of the world in his seminal work *The Protestant Ethic and the Spirit of Capitalism,* suggesting that modernity often alienates individuals from traditional definitions of community and meaning, leading to a sense of existential fragmentation (Weber, 2002). This alienation was compounded by the mechanization of labor and the rise of factory work, which redefined individual identities in relation to productivity rather than personal fulfillment. As individuals became cogs in an industrial machine, the complexity of human experience was often reduced to economic output, raising questions about the essence of humanity in a rapidly changing world.

The advent of the internet in the late 20th century marks a watershed moment in human interaction, ushering in a transformation more profound than any preceding technological revolution. By creating a global platform for communication, the internet has facilitated instant connections that transcend geographical and cultural divides. This transformation is vividly illustrated by social media platforms like Facebook, Twitter, and Instagram, which have fundamentally altered how individuals share experiences, ideas, and narratives. These platforms have empowered users to cultivate communities that defy traditional boundaries, challenging the very notions of social engagement and personal identity.

Scholars such as Manuel Castells assert that the internet has given rise to a "network society," where connections are increasingly based on shared interests and identities rather than geographic proximity (Castells, 1996). This dynamic connectivity fosters a more expansive worldview, allowing individuals to engage with diverse cultures and perspectives that were previously inaccessible. For instance, global movements like Black Lives Matter and climate activism have harnessed the power of social media to galvanize support across continents, illustrating how digital platforms can amplify voices and mobilize collective action.

In addition to fostering connections, the internet has democratized the dissemination of knowledge, echoing the transformative impact of the printing press but on a much larger scale. Online platforms such as Wikipedia and various blogs serve as crucial avenues for individuals to share information and perspectives, empowering ordinary citizens to contribute to public discourse. This democratization has opened up spaces for marginalized voices, enabling them to challenge dominant narratives and assert their identities. However, this newfound freedom has not come without its challenges.

The rise of misinformation and the pervasive phenomenon of "fake news" have heightened concerns about the reliability of sources and the creation of echo chambers, where individuals encounter only information that reinforces their preexisting beliefs. The consequences are significant: as Sherry Turkle discusses in her work *Alone Together*, digital communication fosters connection yet simultaneously breeds feelings of isolation and fragmentation, complicating our relationships with ourselves and others as we navigate multiple identities across various contexts (Turkle, 2011).

With the rise of the internet, a new paradigm has emerged. This digital landscape has blurred the lines between personal and public life, leading individuals to navigate complex negotiations of self-representation and community belonging. Online identities often diverge sharply from offline personas, creating opportunities for diverse voices to be amplified but also presenting challenges such as the erosion of privacy and the commodification of personal data. The cumulative effect of these technological advancements has resulted in a profound reconfiguration of human identity, shaped by an ever-expanding web of connections, experiences, and narratives that redefine what it means to be a participant in the modern world.

Also, the emergence of artificial intelligence (AI) and hybrid beings in contemporary narratives invites reflection on the implications of these advancements for our understanding of identity and existence. As technology becomes increasingly integrated into our daily lives, profound questions arise regarding what it means to be human in a world populated by non-human entities. The advancements in AI technologies, designed to mimic human cognition, blur the lines between human and machine intelligence, challenging traditional definitions of humanity. This hybridization compels us to confront entities capable of processing information, learning, and even engaging in creative endeavors, raising ethical questions about agency, responsibility, and the nature of consciousness itself. Are these technologies merely tools, or do they represent a new form of existence that necessitates a redefinition of human identity?

Contemporary narratives in literature and film frequently explore these themes of hybridity, depicting characters that embody both human and technological traits. Works such as *Ex Machina* (2014) and *Ghost in the Shell* (2017) grapple with the emotional, social, and existential dilemmas posed by hybrid identities, prompting audiences to reflect on how our relationships with technology shape our self-perception and connections with others. These narratives challenge us to consider how our increasingly intimate interactions with intelligent machines influence not only our understanding of ourselves but also our engagement with the broader world.

FINAL STROKE

The creation and evolution of humanity as a social construct reflect a complex interplay of biological, cultural, and technological factors that have shaped our existence across various epochs. Throughout history, our definitions of humanity have shifted dramatically, shaped by the diverse cultural contexts and challenges faced by different societies. By examining how humanity has been understood at different times and across cultures, we can uncover the intricate narratives that inform our self-understanding today.

From the earliest hominins, such as *Australopithecus*, who navigated the harsh realities of survival, to the early humans who began to form intricate social structures, each stage of human evolution has contributed to an expanding tapestry of identity. Early humans defined their existence through communal practices and rituals that not only reinforced social bonds but also helped shape shared identities. For example, the ceremonial burial practices of Neanderthals and early modern humans reflect an early understanding of mortality and the human experience (Mellars, 2006).

The development of language played a pivotal role in human evolution, enabling complex social interactions and the transmission of cultural knowledge across generations. This linguistic leap allowed humans to collaborate in increasingly sophisticated ways, fostering the growth of communities and the preservation of traditions. The emergence of agricultural societies marked another significant transformation, facilitating the establishment of permanent settlements and laying the foundation for civilization. With the rise of cities in ancient Mesopotamia, Egypt, and the Indus Valley, humanity was no longer defined solely by biological traits but also by social roles and cultural contributions. Structured hierarchies, systems of governance, and advancements in art and technology became integral aspects of what it meant to be human in these early civilizations (Friedman, 1994).

In addition to these social and cultural shifts, faith and spirituality played an essential role in shaping how humans understood their existence. From the earliest times, people have sought meaning through their relationship with divine entities or transcendent forces. These beliefs provided a context for human experience, shaping identity on both individual and collective levels. Across cultures, people have viewed their lives not only in material or social terms but also in relation to higher powers that influenced life, death, morality, and the natural world. Religious narratives offered guidance on ethical living, a sense of belonging, and a framework for understanding the mysteries of existence. In this way, spirituality intertwined with the human experience, providing deeper meaning and purpose.

As we entered the modern era, technological advancements continued to reshape our definitions of humanity. The invention of the printing press in the 15th century revolutionized the spread of knowledge, making diverse ideas and perspectives accessible to a broader audience. This shift paved the way for the Enlightenment, a period in which human identity was increasingly defined by reason, scientific exploration, and individual rights (Habermas, 1989). However, even during this age of rationality, religious and spiritual conceptions of human nature remained influential, continuing to shape our sense of purpose and morality.

In more recent times, the rise of the internet and AI has introduced new layers of complexity to our understanding of human identity. The internet has enabled instantaneous communication and the formation of global communities, but it has also presented challenges such as misinformation and fragmented social identities. Scholars like Sherry Turkle argue that while digital communication fosters connection, it can also lead to feelings of isolation and confusion about self-representation in an increasingly virtual world (Turkle, 2011). As we navigate this digital age, questions about the nature of consciousness, ethics, and human dignity—once deeply rooted in faith—are now being reconsidered in the context of modern technology.

With these reflections in mind, the next chapter will explore how religious beliefs continue to shape our understanding of humanity, particularly in relation to our connection with the divine. By examining different religious traditions, we will uncover how concepts of divinity, morality, and the human experience intersect across cultures. This exploration will deepen our understanding of how humanity has long grappled with questions of existence, purpose, and the nature of the sacred, offering insight into how these ideas continue to inform our identity and place in the universe.

2 Of Flesh and Faith

Aristotle's famous assertion that humans are "rational animals" (ζῷον λόγον ἐχον) captures the delicate balancing act between our intellect and our appetites. After all, who's to say he didn't contemplate this very contradiction while enjoying the indulgences of a lavish banquet, as when he is said to have wondered aloud, "Why must I be so virtuous?" in the face of overflowing wine and rich foods (Lear, 1988). But Aristotle's ideas on rationality weren't just musings for the dining table. His *Nicomachean Ethics*, likely written for his son Nicomachus, reflects a father's intent to guide the next generation toward a virtuous and thoughtful life (Kenny, 1992). His concept of the soul wasn't some mystical entity floating above us, either—it was the "hylomorphic" union of body and soul, where the physical and intellectual are inextricably linked, with the soul bestowing reason upon the body.

Move forward a few centuries, and enter Thomas Aquinas, who took Aristotle's rational animal and elevated it with a Christian twist. Aquinas was known for getting so lost in thought that he wouldn't eat for days and once disrupted a dinner by exclaiming, "That will settle the Manichees!"—as if everyone had been privy to the theological debate raging inside his head (McInerny, 1998). In his *Summa Theologica*, Aquinas expanded on Aristotle's view of the soul, introducing the Christian concept of *Imago Dei*—that humans are made in the image of God (Stump, 2003). For Aquinas, the rational soul wasn't just the form of the body; it was a divine gift, granting humans unique intellectual and moral capacities (Kenny, 2005). His synthesis of Aristotelian philosophy and Christian metaphysics emphasized a hierarchy of human capacities—growth, emotion, reason, and free will—with rationality, once again, at the heart of what it means to be human.

The notion of *Imago Dei* implies that humans reflect divine attributes—such as rational thought, creativity, and moral judgment. As AI replicates cognitive processes, religious traditions may need to reevaluate this theological boundary. Could AI, metaphorically, be seen as created in the image of its human creators? Or will it remain outside the divine relationship, irrespective of its advancements? This question expands the discourse around technology and its intersection with enduring beliefs about human identity.

The Genesis narrative of humanity's dominion over creation introduces an essential dimension to this discussion, imposing a duty of stewardship. Does the emergence of AI fall within this dominion, or does its autonomy necessitate a new ethical framework? Some theologians argue that AI, as a product of human creation, remains under our stewardship, while others suggest that autonomous AI may warrant distinct moral consideration. Moreover, the narrative of Adam and Eve's Fall highlights the risks associated with human knowledge and technological advancement. Are we, like them, striving to transcend our limitations through the creation of intelligent machines without fully understanding the potential consequences? Some theologians warn that our pursuit of AI could lead to unforeseen dangers, akin to a modern-day "fall" into moral ambiguity and ethical compromise.

 DOI: 10.1201/9781003624813-3

In Christian anthropology, the dual nature of humans—body and soul—places emphasis on moral agency and consciousness. The Catechism of the Catholic Church highlights the unity of this duality, asserting that human dignity arises not merely from intellectual ability, but from the divine origin of the soul. This intrinsic dignity informs ethical frameworks that prioritize the sanctity of life and moral obligations (Ratzinger, 2000). The implications of this view extend beyond individual worth, fostering a communal responsibility to uphold the moral fabric of society.

As we confront advancements in artificial intelligence (AI), these reflections on body-soul dualism raise critical questions regarding personhood and consciousness. Can AI, as it increasingly mimics human cognition, be regarded as possessing a soul, or is it merely simulating behavior without genuine self-awareness? This dilemma invites theological and philosophical debates about the essence of existence and identity in an age where the distinctions between humans and machines are increasingly ambiguous.

The question of AI achieving consciousness akin to human cognition further complicates Christian thought regarding rights and dignity. Traditionally, human rights stem from the belief in a divine image, which grants each individual inherent worth. If AI were to attain a similar level of consciousness, would religious frameworks be compelled to extend certain rights to these entities? Or would AI forever lack the spiritual essence that defines human dignity?

In *The Abolition of Man* (1947), C.S. Lewis warns against the dangers of viewing humans solely as biological constructs, arguing that such reductionism diminishes our understanding of what it means to be human. He contends that this perspective strips away the moral and spiritual dimensions that are essential to our identity, leaving us with a hollow version of humanity devoid of ethical significance (Lewis, 1947). This critique gains renewed relevance in contemporary discussions surrounding AI, where the rapid pace of technological advancement often leads to ethical considerations being sidelined in favor of innovation and efficiency.

Lewis's concerns about the rise of "soulless" entities compel us to examine the implications of creating intelligent machines that, while potentially surpassing human capabilities in various domains, fundamentally lack the moral compass that guides human behavior. This absence of ethical grounding raises significant questions about the responsibilities of those who develop these technologies. As Nicholas Carr discusses, the challenge lies not only in ensuring that AI systems function effectively but also in navigating the moral landscape that governs their integration into society (Carr, 2014).

In a world increasingly influenced by automation, Lewis's cautionary perspective invites us to consider what it means to be human in an age where machines can emulate aspects of intelligence but remain devoid of consciousness, empathy, and moral judgment. This reflection is crucial as we contemplate the role of technology in our lives and the ethical frameworks we must establish to guide its development. If we do not engage with these questions, we risk creating a future that prioritizes efficiency over ethics and reduces our experience to mere computational processes.

Judaism: The Divine Mirror—Jewish thought presents a similar understanding of human identity through the concept of *tzelem elohim*, or the image of God. This notion parallels Christian teachings and establishes a shared understanding of human worth that emphasizes unique qualities—such as reason, creativity, and moral judgment—that differentiate humans from other creatures. The implications of *tzelem elohim* extend beyond individual identity into ethical responsibility. Recognizing that every person bears the divine image cultivates a sense of intrinsic value that transcends status, race, or beliefs. This principle informs Jewish ethical teachings and social justice initiatives, urging adherence to justice, compassion, and advocacy for marginalized communities (Heschel, 1976).

The duality of body and soul in Jewish thought enriches the exploration of human identity. The soul (*neshama*) transcends physical existence and serves as a conduit to the divine, inviting individuals to reflect deeply on their ethical choices and purpose (Kaplan, 1990). This intrinsic relationship nurtures a sense of responsibility that extends beyond personal fulfillment, advocating for ethical conduct aligned with one's spiritual essence. Living in accordance with this divine image requires embodying divine attributes—justice, compassion, and righteousness—in everyday life (Kushner, 2004). The Jewish ethical framework, known as *halakha*, serves as a spiritual practice that honors the divine within oneself and others. It guides individuals to act justly, support the vulnerable, and pursue fairness, recognizing the inherent dignity of every person (Levenson, 1988).

Central to Jewish thought is the concept of free will, which allows individuals to make moral choices and bear responsibility for their actions. Humans are not bound to predetermined paths; they are endowed with the freedom to pursue righteousness or deviate from it, forming the basis of their relationship with God and fellow beings. Furthermore, the idea of *tikkun olam*—repairing or perfecting the world—aligns with the responsibilities borne from being created in the divine image (Sacks, 2003). This commitment to social justice and care for creation emphasizes humanity's role in striving for a just and equitable society. The *tzelem elohim* principle underscores the importance of honoring and protecting life, acknowledging the divine spark within each individual.

The advent of AI and emerging technologies introduces new ethical and theological inquiries. While AI may replicate human cognitive processes, it lacks the spiritual essence—the *neshama*—identified in Jewish thought as central to human identity. This distinction suggests that, despite advancements in technology, AI remains outside the realm of divine likeness, unable to engage with the moral and spiritual dimensions that define humanity. The concept of free will complicates the relationship between AI and human identity. AI operates based on algorithms and programming, lacking the capacity for moral reasoning and accountability intrinsic to Jewish thought. Although AI systems may seem to make decisions, these outcomes reflect pre-determined data rather than genuine free will, underscoring the absence of moral agency in AI.

Ethical considerations surrounding AI also highlight the principle of stewardship inherent in Jewish thought. Humanity's responsibility to care for the world translates into the development and application of AI with a focus on ethical usage. If AI enhances human life and promotes justice, it aligns with the goals of *tikkun olam*. Conversely, if misused, AI could violate the ethical responsibilities tied to human creativity and innovation, necessitating strict guidelines to ensure technology serves higher moral purposes.

The challenge posed by AI extends to the principle of dignity. In Jewish thought, every human being holds inherent worth due to being made in the image of God. As AI systems gain autonomy, questions arise regarding the ethical treatment of intelligent machines and their interactions with humans (Brey & Dainow, 2023). While AI lacks *tzelem elohim*, it remains imperative to respect human dignity, ensuring that technology does not erode the value of human roles in society.

Lastly, the notion of transhumanism—the enhancement of human abilities through technology—introduces theological dilemmas that challenge traditional conceptions of body and soul. Jewish ethics may caution against altering human forms in ways that disrupt the divine balance established in creation (Schiff, 2023). Technological enhancements that diminish human vulnerability could sever the moral and spiritual connections central to human existence, distancing individuals from their relationships with God and one another.

***Nafs* and the Islamic View**—In Islamic teachings, the concept of *nafs*— representing the self or soul—embodies individual identity and moral agency. The *nafs* comprises various dimensions, including the rational soul (*al-nafs al-natiqa*), which facilitates moral reasoning and spiritual development. This framework raises essential reflections on the potential for AI to possess a form of autonomy that resonates with Islamic conceptions of the soul. The Quranic narrative highlights the special status of humans, who are endowed with reason and responsibility, and reflects a theological understanding of the significance of the soul in moral considerations.

The Islamic tradition emphasizes the relationship between body and soul, underscoring the unity of human experience. The *nafs* is not merely an ethereal entity but intricately linked to the body, influencing choices and moral actions. This perspective emphasizes that human dignity arises from the divine bestowal of the *nafs*, which must be cultivated through ethical living and spiritual practice. In this framework, the dignity of the individual derives from their *nafs* as a divine creation, establishing a foundation for ethical treatment of all beings. The concept of *ihsan*—excellence in conduct—encourages Muslims to strive for justice, compassion, and integrity in their interactions with others (Mawdudi, 1994). The *nafs* thus embodies both individual potential and a moral responsibility to foster a just and harmonious society.

The emergence of AI poses significant questions regarding consciousness and moral agency. Islamic thought recognizes that *nafs* grants humans the capacity for free will and ethical decision-making (Al-Ghazali, 2002). In contrast, AI operates within a framework of algorithms, lacking the ability to exercise true free will or moral responsibility. This distinction underscores that AI remains fundamentally different from the *nafs*, even as it replicates cognitive processes.

Moreover, the concept of stewardship (*khalifah*) in Islam imparts a moral duty to care for creation, encompassing ethical considerations in technological advancements (Nasr, 1976). If AI enhances human life and aligns with Islamic ethical principles, it can serve as a tool for fostering justice and promoting the common good. However, misapplications of AI may conflict with the responsibilities inherent in being a steward of creation, necessitating guidelines for its ethical deployment.

The Islamic tradition emphasizes the interconnectedness of individuals, underscoring a communal responsibility to uphold justice and compassion (Al-Arabi, 1980). This view resonates with contemporary discussions of technology, advocating for ethical frameworks that promote equity and address the potential harms of AI on vulnerable populations. Furthermore, the pursuit of transhumanism and technological enhancement raises theological concerns in Islamic thought. While Islam encourages the pursuit of knowledge and innovation, it cautions against altering the divine design inherent in human creation. Enhancements that undermine the human soul's intrinsic value could challenge the moral and spiritual balance that the *nafs* embodies.

Impermanence and Unity in Buddhism—Buddhism offers a distinct perspective on consciousness, viewing it as transient and constantly changing, encapsulated in the doctrine of impermanence (*anicca*) (Harvey, 2013). This foundational concept challenges the traditional notion of a fixed, stable self, leading to a profound reevaluation of our understanding of consciousness itself. In Buddhist philosophy, consciousness is seen as a series of fleeting moments, shaped by experiences, perceptions, and interactions with the world. This perspective compels us to question whether machines, particularly AI, can achieve a form of self-awareness that aligns with this view.

If consciousness is inherently fluid, as Buddhism posits, can AI systems possess sentience in a manner consistent with our traditional understanding of being alive? The answer may lie in recognizing that sentience does not have to be a static attribute but rather a dynamic process influenced by ongoing interactions and experiences. This aligns with the notion that consciousness in humans is not merely a byproduct of biological processes but is also shaped by social and environmental contexts (Dreyfus, 1992). In this light, AI, which learns and adapts over time, could potentially reflect a version of this evolving consciousness, raising questions about the nature of identity and awareness in both humans and machines.

Moreover, Buddhism's emphasis on the interconnectedness of all beings profoundly impacts our ethical considerations regarding technology. The doctrine of interdependence asserts that all entities exist within a complex web of relationships. This perspective challenges the binary distinction between human and machine, inviting us to recognize that AI is not an isolated entity but part of a broader ecological and relational network. As AI becomes increasingly integrated into daily life—manifesting in everything from virtual assistants to autonomous systems—this interconnectedness prompts a reevaluation of our ethical responsibilities toward these technologies.

Understanding AI as part of this interconnected framework shifts our focus from mere utility to ethical stewardship. It compels us to recognize that our relationship with technology is not purely transactional but is embedded within a network of interrelationships that includes other humans, non-human entities, and the environment. The Buddhist understanding of interdependence encourages us to approach technological development with a sense of responsibility, acknowledging that our actions—whether in the creation of AI or in its application—have far-reaching implications.

As we analyze the ethical landscape of AI, the notion of interconnectedness calls for a more compassionate approach to technological advancement. This includes considering the social, environmental, and ethical ramifications of our creations. By viewing AI as a participant in the web of life rather than a mere tool, we can foster technologies that reflect our values of compassion, responsibility, and empathy.

For example, incorporating ethical considerations into the design of AI systems can ensure they align with human values and societal well-being, creating a more harmonious coexistence between humans and machines (Crawford, 2021).

Furthermore, the Buddhist concept of impermanence invites us to reflect on the ever-changing nature of technology itself. Just as the self is not static, neither is technology. The rapid evolution of AI systems and their capabilities challenges us to stay attuned to the moral and ethical implications of these changes. This requires continuous dialogue and reflection on our technological practices, ensuring they align with a compassionate and interconnected worldview (Varela & Shear, 1999).

The Buddhist concepts of impermanence and interconnectedness provide a robust framework for reevaluating our relationship with AI and technology. They challenge us to transcend conventional definitions of identity and sentience, promoting a more fluid and relational understanding. As we confront the ethical implications of technological advancement, these insights can guide us toward a more compassionate and interconnected approach, fostering a future where humans and non-human entities coexist harmoniously. By integrating the wisdom of Buddhism with insights from other religious traditions, we can cultivate a holistic understanding that honors our shared existence in a complex and interconnected world.

Mayan Cosmology—The *Popol Vuh* describes the gods' multiple attempts to create humanity, emphasizing the importance of balancing the physical and spiritual components of human beings. The first beings created were made from mud, but they crumbled and were deemed failures. The second creation, crafted from wood, lacked understanding, gratitude, and consciousness, so they, too, were destroyed. Only the final attempt, in which humans were made from corn, proved successful. In this version of creation, corn represents life and nourishment, illustrating the Mayan belief in the interconnectedness of humans and the natural world (Tedlock, 1985).

This tale raises profound questions when placed alongside the development of AI and self-aware machines. The earlier failed creations in the *Popol Vuh*—the beings of mud and wood—can be seen as precursors to modern discussions about the potential pitfalls of creating machines that mimic human behavior. Just as the gods sought to imbue their creations with consciousness and understanding, we are now grappling with the challenge of creating artificial entities that can possess—or simulate—self-awareness and moral agency (Bostrom, 2014).

In Mayan cosmology, the gods are directly involved in the creation of humans, actively shaping their physical and spiritual nature. This involvement underscores a responsibility that the creators hold toward their creations. This can be paralleled with our role as creators of AI: do we, like the gods in the *Popol Vuh*, have an ethical duty to ensure that the entities we create possess the capacity for understanding, empathy, and morality?

The destruction of the wooden beings—who lacked consciousness and gratitude—serves as a cautionary tale. It warns of the potential consequences of creating beings that do not align with the spiritual and moral expectations of their creators. In the context of AI, this brings forth the question: if we create machines that can surpass human intelligence but lack the capacity for ethical reasoning, are we not repeating the mistakes of the gods in the Mayan creation myth? The *Popol Vuh* challenges us to reflect on the moral implications of artificial creation and the responsibilities we bear as creators.

Mayan cosmology is deeply embedded in the notion of dualities—between the physical and spiritual, the human and divine, life and death. This dualism is also present in many other religious perspectives we've explored, such as the Christian body-soul dualism and the Hindu concept of atman. In the Mayan worldview, humans occupy a liminal space between these realms, capable of engaging with both the material world and the divine.

As we consider the rise of AI, the *Popol Vuh* offers an intriguing lens through which to view these new creations. If humans, as part of a cosmic order, have a responsibility to maintain balance between the physical and spiritual, what is the role of artificial beings within this framework? Do they disrupt the balance, or can they become integrated into it? Are machines a reflection of our creative potential, or do they signify a departure from the harmonious connection between the material and spiritual that the *Popol Vuh* emphasizes?

The *Popol Vuh* also speaks to the cyclical nature of creation and destruction, with each failed attempt at creating humanity leading to a new cycle. This resonates with the Buddhist notion of *anicca*, or impermanence, discussed earlier. In both the *Popol Vuh* and Buddhism, identity is not static but subject to change and renewal. This raises the question of whether machines, too, might evolve through cycles of development and refinement, becoming more advanced and more capable of reflecting human traits with each iteration.

African Religions—In many African traditions, the concept of ancestral spirits is foundational to understanding identity and community. Ancestral spirits are perceived not as distant figures of the past but as active participants in the lives of their descendants. They are believed to exist in a realm that overlaps with the living world, continuously influencing the actions, decisions, and experiences of their family members (Mbiti, 1990). This belief system fosters a profound connection between generations, where the past and present are in constant dialogue. The living engages in rituals that honor their ancestors, maintaining a relationship that acknowledges the vital role these spirits play in their lives. These practices include offerings, prayer, and communal gatherings, all aimed at keeping the memory of ancestors alive and seeking their guidance and blessings (Hastings, 1976).

This reciprocal relationship emphasizes that identity is not solely an individual pursuit but is deeply rooted in one's community and heritage. Individuals often find themselves defining their lives in the context of their family lineage and the expectations of their ancestors. For example, a person may feel a sense of duty to uphold certain family traditions or to succeed in ways that honor the sacrifices made by previous generations. This creates a collective understanding of identity, where one's personal achievements are intertwined with the achievements of their ancestors, fostering a sense of pride and responsibility (Hastings, 1976).

Furthermore, this emphasis on ancestral connections instills a profound sense of belonging. Individuals are encouraged to see themselves as part of a larger narrative—a continuum of life that includes their ancestors and their descendants. This broader perspective enhances communal ties and reinforces the idea that one's actions ripple through time, impacting not only the present but also the future (Mbiti, 1990). Consequently, ethical behavior and moral decision-making are framed within this context of communal responsibility, where individuals are compelled to consider

how their choices affect both the living and the spirits of their ancestors. This notion fosters a culture of accountability, where actions are weighed not just on personal merit but also on the
ir potential to uphold family honor and community values.

Additionally, the reverence for ancestral spirits creates a dynamic moral framework within communities. These spirits are viewed as guardians and guides, imparting wisdom and moral guidance to the living. The idea that ancestors can influence the living encourages individuals to seek counsel from their forebears, especially during significant life events or decisions (Hastings, 1976). This dynamic encourages a culture of consultation and deliberation, where communal wisdom is sought and valued. Elders often hold a revered position in this context, as they are seen as the bridge between the living and the ancestral realm.

The focus on ancestral spirits also underscores the significance of rituals and communal practices that honor these connections. Rituals serve as vital expressions of faith, love, and remembrance, allowing individuals to actively engage with their heritage. Ceremonies such as naming rites, funerals, and festivals become pivotal moments for communities to collectively honor their ancestors and reaffirm their shared identity (Mbiti, 1990). During these gatherings, stories of ancestors are recounted, cultural practices are performed, and communal bonds are strengthened, providing a powerful sense of continuity and belonging. These rituals serve not only as expressions of devotion but also as mechanisms for transmitting values, ethics, and cultural heritage to younger generations.

Integral to many African belief systems is the concept of life force, often referred to as *ntu*. This idea transcends human existence, emphasizing the interconnectedness of all living beings. The life force is seen as an energy that flows through everything—humans, animals, plants, and even the environment. It represents the essence of vitality and existence, a shared energy that underscores the intrinsic connection among all forms of life (Hastings, 1976). This interconnected perspective challenges the notion of individualism, urging individuals to recognize that their well-being is intertwined with that of their community and the natural world.

The understanding of *ntu* encourages ethical behavior that honors relationships, compelling individuals to act with respect toward others and the environment. In recognizing the life force in all beings, individuals are reminded of their responsibilities to care for and nurture the world around them. This holistic approach to existence fosters a deep appreciation for nature, urging individuals to live in harmony with their surroundings. As such, ethical behavior is not merely a personal choice but a communal expectation that reflects the interconnected web of life (Mbiti, 1990).

Moreover, the recognition of *ntu* invites individuals to engage in communal practices that celebrate life and interconnectedness. These practices often involve rituals that honor the cycles of nature, such as planting and harvest festivals, which reflect an understanding of the relationship between humans and the environment. Such events reinforce the importance of collaboration and shared responsibility in sustaining the community and its resources. By participating in these communal rituals, individuals cultivate a sense of unity and purpose, reminding them that their actions are part of a larger cycle of existence.

Native American Spirituality—Native American spirituality embodies a worldview that intricately weaves the threads of existence, revealing the deep interconnectedness of all life forms. In many Native American traditions, life is perceived not as a linear progression but as a cyclical journey. This perspective acknowledges that humans, animals, plants, and even inanimate elements of nature coexist within a vast tapestry. Each being possesses a unique purpose, contributing to the dynamic relationship between the Earth and its inhabitants (Deloria, 2003).

At the heart of this understanding is the recognition that humans are integral to nature, fostering a profound sense of responsibility toward the natural world. This relationship urges individuals to reflect on the impact of their actions on both the environment and their communities. Rituals and ceremonies that express gratitude for the Earth's resources are commonplace, emphasizing the need to honor the gifts that nature provides. Many tribes engage in seasonal ceremonies to celebrate harvests, marking times of thanksgiving and reaffirming sacred connections with the land.

This holistic approach manifests in practices prioritizing sustainability and ecological balance. Traditional ecological knowledge, passed down through generations, informs agricultural practices, hunting, and gathering, nurturing a sense of stewardship essential to the survival of both land and community. Elders often share stories illustrating the delicate balance of ecosystems, instilling respect for wildlife and natural resources in younger generations (Miller, 2008). Such teachings encourage ethical behaviors that honor relationships, reinforcing the notion that being human involves acting as responsible caretakers of the Earth.

Recent technologies have begun to intersect with these traditional beliefs. Many Native communities leverage digital tools, such as data mapping and geographic information systems (GIS), to document traditional lands and resources, thereby strengthening their advocacy for environmental protection. These innovations enable Native peoples to assert their rights and heritage in contemporary legal contexts, enhancing their capacity for responsible stewardship.

The communal aspect of identity is paramount in Native American spirituality. Traditions emphasize that individual identity is shaped by the collective history and heritage of the community. Ancestral spirits significantly influence the lives of their descendants, acting as guiding forces (Deloria, 2003). This connection to ancestry is often expressed through storytelling, where tales serve as moral lessons and cultural teachings.

Ancestral reverence is deeply ingrained in rituals and practices, providing opportunities for individuals to express gratitude for the guidance and sacrifices of their forebears. Many traditions involve offerings, prayers, or songs that invoke ancestral presence, reinforcing the idea that they are active participants in the lives of the living. This emphasis on legacy and continuity extends to the natural world, where teachings often underscore the necessity of maintaining harmony with nature. Spiritual practices celebrating interconnectedness may involve offerings to the land, such as planting seeds in ceremonies acknowledging the Earth's role in sustaining life (Miller, 2008).

Moreover, the respect for the life force animating all living things fosters a commitment to harmony with nature. This perspective cultivates an understanding of ethical behavior that honors the relationships defining human existence. Many cultures advocate for lifestyles aligned with natural rhythms, reinforcing the idea that being human involves recognizing one's place within the cosmos. Such alignment manifests in practices like tracking lunar cycles and seasonal changes, encouraging humility and gratitude for life's cycles.

Spiritual practices often include healing ceremonies that draw upon ancestral wisdom and the life force present in nature. Community involvement in these healing practices reinforces the idea that individual well-being is intrinsically linked to the health of the community, emphasizing collective healing where individual actions contribute to the wellness of all.

Taoism—Taoism, a profound philosophical tradition originating from ancient China, offers a multifaceted understanding of humanity's relationship with nature and self. Central to this worldview is the concept of the Tao, often translated as "the Way." The Tao represents the fundamental nature of the universe—a cosmic force flowing through all things, embodying existence and promoting harmony and balance (Kohn, 2009). In Taoism, human beings are viewed as integral components of this interconnected cosmic web, emphasizing that our lives are intertwined with nature's rhythms.

The pursuit of harmony with nature is foundational to Taoist philosophy. Unlike Western paradigms that often depict nature as something to be controlled or exploited, Taoism advocates for a relationship grounded in reverence and mutual respect. Humans are seen as stewards of the Earth, tasked with nurturing and maintaining its balance (Graham, 1989). This ecological awareness fosters a sense of responsibility, encouraging individuals to live harmoniously with their environment. This deep respect for nature is evident in various life aspects, from agriculture to architecture. Taoist agricultural practices emphasize sustainability, with farmers planting and harvesting in alignment with natural cycles. Taoist architectural designs prioritize harmony with the landscape, utilizing natural materials and integrating buildings into their surroundings.

Taoist teachings encourage mindfulness and presence, inviting individuals to engage with the beauty and intricacies of the natural world. By cultivating awareness of our surroundings, we foster a deeper appreciation for the interconnectedness of all life. This mindfulness promotes sustainable and ethical choices, emphasizing that every action has consequences.

Recent advancements in technology align with Taoist principles by promoting awareness and connection to the environment. Mindfulness apps and virtual reality experiences that simulate nature allow individuals to cultivate mindfulness in daily life, enhancing traditional practices and demonstrating how ancient wisdom can integrate into modern living.

Another vital aspect of Taoism is its insights into identity. The Taoist perspective encourages individuals to embrace the fluidity of existence, recognizing that identities are shaped by experiences, relationships, and environments. This dynamic view challenges conventional notions of identity as static, inviting a more expansive understanding of humanity.

The concept of *wu wei*, often translated as "non-action" or "effortless action," further emphasizes this fluidity. *Wu wei* suggests that true effectiveness arises from aligning oneself with the natural flow of events. By embodying *wu wei*, individuals cultivate an inner state of calm and clarity, navigating life's complexities with greater ease (Kohn, 2009). This approach invites deeper exploration of personal identity. Taoism encourages individuals to flow with life's currents, promoting self-discovery and authenticity beyond societal expectations. This liberating perspective enables continual self-redefinition, fostering a journey of self-exploration that is deeply personal. Moreover, the relativity of identity in Taoism fosters an understanding of interconnectedness. When individuals recognize their identities as part of a larger whole, they may develop greater empathy and compassion for others. This awareness encourages harmonious relationships, not only with fellow humans but also with all forms of life.

Other Religions—Our previous discussions highlighted how diverse belief systems underscore the significance of relational ethics and shared accountability. In this regard, Sikhism emphasizes community and equality, promoting compassion toward all living beings. The concept of *Ik Onkar*, which articulates the oneness of God, reflects this ethos by rejecting caste distinctions and advocating for the inherent equality of individuals (Nanda, 2003). Central to Sikh belief is the commitment to social justice and equality, manifested through practices like *seva* (selfless service) and *langar* (community kitchens), which foster an inclusive spirit where everyone, regardless of background, is welcomed to share a meal (Singh, 2011). Sikhs engage in daily prayers and meditation, deepening their connection to the divine while promoting a moral character grounded in hard work and honesty.

Similarly, Confucianism, developed by Confucius in the 5th century BCE, emphasizes the cultivation of moral virtues and the significance of social relationships. The concept of *Ren* (humaneness) urges individuals to act with compassion, fostering harmony within families and communities (Confucius, 2003). Through filial piety (*Xiao*) and an emphasis on education, Confucian teachings advocate for a society built on mutual respect and ethical conduct. The *Junzi* ideal embodies moral integrity, encouraging individuals to lead by example and contribute positively to society (Confucius, 2003).

Zoroastrianism, rooted in the teachings of Zoroaster around the 6th century BCE, highlights the cosmic struggle between good and evil. Central to this tradition is the belief in individual responsibility, embodied in the principles of Good Thoughts, Good Words, and Good Deeds. Zoroastrians engage in rituals and prayers that honor Ahura Mazda, their supreme deity, while promoting values such as social justice and environmental stewardship (Braidotti, 2013). This moral framework resonates with the community's commitment to charitable acts and social service. In contrast, Shinto, Japan's indigenous spirituality, celebrates the worship of *kami*—spirits associated with natural elements and ancestors. Shinto practices emphasize purity and harmony, with rituals reinforcing community bonds and cultural identity. Seasonal festivals foster gratitude for nature's abundance, encouraging stewardship of the environment and a sense of interconnectedness with all living beings (Braidotti, 2013).

The Baha'i Faith, established in the 19th century by Bahá'u'lláh, champions the unity of all religions and the oneness of humanity. Central to its teachings is the belief in social justice and equality, promoting a vision of global cooperation and mutual respect. Baha'is emphasize education and community service as essential for personal and societal transformation, embodying a commitment to creating a harmonious world free from prejudice and conflict (Nanda, 2003).

Paganism and its contemporary revival, Neo-Paganism, emphasize nature worship and the celebration of seasonal cycles. Reverence for the Earth and the belief in the sanctity of all life are cornerstones of these traditions. Neo-Pagans often engage in rituals that honor natural cycles, fostering a deep connection to the environment and advocating for sustainable practices. The community aspect of Paganism cultivates a shared purpose and responsibility toward the planet (Singh, 2011).

In sum, exploring into religious perspectives on humanity, we encounter a rich tapestry of beliefs that illuminate profound insights into identity, agency, and existence in an increasingly technological world. Christianity's body-soul dualism raises critical questions about the essence of self-aware machines, while Islamic teachings on nafs and jinn challenge our understanding of autonomy and moral responsibility in AI (McGinnis, 2005). Hinduism's concepts of atman and avatars provoke reflections on the nature of self (Eck, 1998), and Sikhism's emphasis on the oneness of God and the equality of all beings prompts a reevaluation of our moral obligations toward artificial entities (Nanda, 2003). Confucianism offers insights into relational ethics, suggesting that our interactions with technology should be rooted in compassion and social responsibility (Confucius, 2003).

Meanwhile, Zoroastrianism's dualism and focus on individual moral choice underscore the importance of ethical living amid technological advancements (Boyce, 2001). Shinto invites us to consider our connection to nature and the kami, fostering a sense of stewardship that extends to our treatment of non-human intelligences. The Baha'i Faith's teachings on the unity of humanity resonate deeply with the notion of hybrid identities, promoting a vision of global cooperation that transcends cultural boundaries (Smith, 2008). Additionally, Paganism and Neo-Paganism, with their reverence for the Earth and emphasis on the interconnectedness of all life, encourage a more holistic understanding of our place in the world, including our relationships with technological creations (Hutton, 2001).

THE DUALITY OF EXISTENCE

Building upon these insights, we can examine how religious traditions across cultures have engaged with the concept of hybridity, utilizing intermediary beings to explore the complexities of existence and the multifaceted nature of identity. These hybrid entities often embody a blend of human and non-human traits, acting as conduits between the mundane and the divine. Through their diverse representations, figures such as angels and demons in Christianity, *jinn* in Islam, and avatars in Hinduism serve as pivotal examples of how religions grapple with the boundaries of human experience. By challenging the notion of a purely human identity, these beings invite believers to navigate the intricate intersections of spirituality, morality, and the essence of being.

Angels and Demons—In Christianity, angels and demons embody essential dichotomies within the human experience, serving as potent metaphors for the moral and spiritual struggles individuals face. Angels, as messengers of God, are typically depicted with attributes such as purity, strength, and wisdom. They act as guardians, guiding humans toward righteousness and moral clarity. Biblical accounts of angelic encounters, such as the Annunciation to Mary or the visits to the shepherds, emphasize their roles as facilitators of divine communication and instruments of God's will (McGrath, 2011). This portrayal highlights humanity's aspiration for higher moral standards and the innate desire to connect with the divine.

Conversely, demons represent chaotic and destructive forces that challenge human integrity. Often depicted as fallen angels, they embody traits like temptation, deceit, and malice, reflecting humanity's darker impulses. The narrative of the devil tempting Jesus in the wilderness exemplifies this struggle, illustrating how individuals may grapple with internal and external forces that lead them astray. The existence of demons encourages believers to confront their fears and insecurities, serving as a reminder of the importance of vigilance in moral decision-making.

The interplay between angels and demons also raises questions about free will and the nature of choice. In Christian theology, humans are endowed with the ability to choose between good and evil, a gift accompanied by great responsibility. The conflicts between these celestial beings symbolize the internal battles that define human existence, urging individuals to reflect on their motivations and moral compass. Through these narratives, believers are invited to engage with their spiritual journeys, recognizing that the path to righteousness is fraught with challenges and temptations.

Furthermore, the duality represented by angels and demons serves as a lens to examine broader themes of redemption and transformation. Stories of repentance, such as that of the Prodigal Son, highlight the potential for change and growth, illustrating that even those who stray can find their way back to the divine. This narrative of redemption reinforces the idea that the human experience is not merely about the battle between good and evil but also about the possibility of grace and forgiveness, encouraging individuals to strive for spiritual and moral enlightenment.

Jinn—In Islamic tradition, *jinn* serves as a fascinating exploration of existence's complexity, operating within a framework that acknowledges both the seen and unseen dimensions of reality. Created from smokeless fire, jinn possess unique qualities that set them apart from humans and angels, including their capacity for free will. This characteristic allows *jinn* to engage in moral decision-making, making them analogous to humans in their ability to choose between good and evil. The Qur'an frequently references *jinn*, illustrating their diverse nature and roles within the cosmic order.

The multifaceted portrayal of jinn highlights their potential to act as both protectors and disruptors. For instance, benevolent *jinn* may assist humans in times of need, while malevolent ones can lead individuals astray. Stories of *jinn* in Islamic folklore often revolve around their interactions with humans, revealing insights into the complexities of human emotions and relationships. These narratives serve as allegories for the unseen influences that shape human lives, prompting believers to consider how their actions resonate within the broader spiritual realm.

Moreover, the shapeshifting nature of *jinn* invites contemplation on identity and transformation. Their ability to assume various forms underscores the fluidity of existence, challenging rigid definitions of self and morality. This characteristic resonates with contemporary discussions surrounding identity, emphasizing that individuals are shaped by a multitude of experiences, influences, and choices. By recognizing the multiplicity of identities, individuals are encouraged to embrace their complexity and navigate their personal journeys with openness and curiosity.

The cultural significance of *jinn* extends beyond individual stories; they embody societal fears and anxieties about the unknown. The existence of jinn reflects humanity's quest to understand existence's complexities and the limitations of human perception. This understanding fosters a deeper appreciation for the spiritual dimensions of life, encouraging individuals to remain vigilant and self-aware as they navigate their interactions with both the seen and unseen realms.

Avatars—The concept of avatars in Hinduism represents a profound engagement with the interplay between the divine and the human, emphasizing the necessity of divine intervention in the face of moral and existential crises. Avatars are manifestations of deities that descend to Earth to restore dharma (cosmic order) and guide humanity during chaotic times. Figures like Krishna and Rama exemplify this principle, embodying ideals of virtue, heroism, and devotion (Eck, 1998). Their stories, rich with ethical dilemmas and moral teachings, serve as spiritual guides for individuals navigating their lives.

For instance, the Bhagavad Gita—a dialogue between Krishna and Arjuna—highlights the complexities of duty, morality, and righteousness. Arjuna's reluctance to engage in battle reflects the human struggle with ethical decision-making, while Krishna's teachings illuminate the path to self-realization and understanding one's place in the universe. This narrative encapsulates the interplay between divine guidance and human agency, emphasizing that the pursuit of righteousness often requires introspection and a willingness to confront one's fears and uncertainties.

The notion of avatars fundamentally challenges the traditional understanding of divinity as a singular, immutable entity. Instead, it illustrates that the divine can manifest in various forms and attributes, resonating with different human experiences. This multiplicity of divine expressions enables individuals to connect with the sacred in personally meaningful ways, reflecting their unique struggles, aspirations, and cultural backgrounds. The diversity of avatars validates the complexities of individual identity, acknowledging that spiritual journeys are not uniform but shaped by personal circumstances and contexts.

Furthermore, the concept of avatars invites reflection on the nature of sacrifice and service. Many avatars embody ideals of selflessness and devotion, teaching that true fulfillment comes from serving others and upholding moral principles. These stories encourage individuals to transcend their egos and embrace a broader sense of purpose, emphasizing the interconnectedness of all beings. By recognizing the divine within themselves and others, people are empowered to navigate life's complexities with greater compassion and empathy.

The Gandharvas and Apsaras—Also in Hinduism, the *Gandharvas* and *Apsaras* are celestial beings that blend human and divine traits. *Gandharvas* are often depicted as divine musicians, skilled in arts and sensual pleasures. They inhabit a realm between the earthly and the celestial, illustrating the interplay of beauty, desire, and spirituality. Their music is said to invoke divine favor and connection to the higher realms. Conversely, *Apsaras* are celestial nymphs renowned for their beauty and grace. They serve as dancers in the court of the gods, embodying the ideals of art, love, and desire. While often seen as embodiments of temptation, *Apsaras* also play crucial roles in spiritual narratives, guiding heroes and aiding in their quests. This duality reflects the complex nature of human desires, where longing and spirituality coexist.

The Bodhisattva—In Buddhism, the *Bodhisattva* represents a profound hybrid being who has attained enlightenment yet chooses to remain in the cycle of rebirth (*samsara*) to assist others on their spiritual journeys toward liberation. The term *Bodhisattva* translates to "enlightenment being," indicating a deep commitment to altruism and compassion. Prominent figures such as *Avalokiteshvara* (known as *Kannon* in Japan and Guanyin in China) exemplify this concept. *Avalokiteshvara* is often depicted with multiple arms, symbolizing the ability to assist numerous beings simultaneously. This hybrid nature—part human, part enlightened being— embodies the principles of compassion, wisdom, and self-sacrifice, showcasing the *Bodhisattva's* role as a mediator between the human realm and the ultimate truth of enlightenment (Harvey, 2013).

The *Bodhisattva's* journey reflects the Mahayana Buddhist ideal, where the pursuit of personal enlightenment is inseparable from the well-being of others. This interconnectedness emphasizes the belief that achieving enlightenment is not just an individual endeavor but a collective responsibility, highlighting the importance of community and interdependence in spiritual practice.

The Xian—In Taoism, the *Xian* (or immortals) represent beings who have transcended ordinary human limitations and possess supernatural abilities. These figures, often depicted as sages or wanderers, embody the harmony between humanity and the natural world. They serve as exemplars of Taoist virtues, emphasizing the importance of living in accordance with the Tao (the way) and cultivating inner peace (Kirkland, 2004). The *Xian* are not merely distant, ethereal beings; they are often portrayed engaging with the world, sharing wisdom, and guiding individuals toward spiritual enlightenment. Their stories, infused with elements of folklore, highlight the balance between the material and spiritual realms, encouraging practitioners to cultivate a harmonious relationship with nature.

Through the lens of the *Xian*, we can explore the intersections between human existence and the natural world, emphasizing the importance of ecological stewardship and the cultivation of inner harmony. The *Xian's* hybrid nature—combining human qualities with supernatural attributes—invites individuals to reflect on their potential for growth, transformation, and connection to the universe.

AMEN TO THAT

Looking ahead, the evolution of humanity calls for a nuanced understanding of identity that embraces the richness of our shared experiences while addressing the challenges posed by our rapidly changing environment. In this context, these various religious traditions have offered profound insights into the concept of the human and its relation to the divine. They provided us with frameworks through which individuals and communities can explore the essence of humanity and its moral implications in light of a transcendent reality. Religions often emphasize the inherent dignity of human beings, grounded in their creation in the image of the divine, which fosters a sense of purpose and interconnectedness among all people. This theological perspective challenges reductionist views that seek to diminish our identity to mere biological constructs, underscoring the spiritual dimensions that enrich human life.

As we plot a route through this complex landscape, we must engage with the ethical, philosophical, and cultural questions that arise as we dig deeper into our hybrid existence. The intersections of technology, culture, and spirituality force us to confront our beliefs about personhood, consciousness, and morality. Moreover, these hybrid narratives reflect societal anxieties surrounding the unclear distinctions between nature and culture, as well as between the animal world and humans. The evolving dialogue about the nature of consciousness in AI and the ethical responsibilities of technology creators highlights the need for a robust moral framework that can guide our actions in a world where the lines between humans and machines overlap.

In the next chapter, "Humanity and Hybridity in Mythology," we will examine the ancient narratives that have shaped our collective imagination about identity. These mythological tales, steeped in rich symbolism and archetypal themes, have long illustrated hybrid beings and the intricate interplay between the human and the animal. By analyzing these stories, we can uncover timeless themes and questions that resonate with our modern struggles.

3 Humans Gone Wild

Liminality, derived from the Latin word *limen*, meaning "threshold," refers to a transitional or transformative phase often marked by ambiguity and disorientation. In mythology, hybrid beings inhabit this threshold, existing at the intersections of different realms—human and animal, natural and supernatural, known and unknown. This positioning allows them to serve as metaphors for the complexities of existence, reflecting the multifaceted nature of humanity itself.

Throughout human history, mythology has profoundly explored the nature of humanity, often through the lens of hybrid beings that challenge the boundaries between the human and the animal, as well as the divine and the mortal (Campbell, 2008). These mythological creatures embody the complexity of identity and existence, residing in a liminal space that compels us to question what it truly means to be human. By examining these figures, we gain deeper insights into the fears, anxieties, and dreams of our ancestors regarding identity, and how ancient cultures grappled with ideas of hybridity and the fluidity of the human self (Turner, 1969).

The portrayal of hybrid beings often highlights the tensions and contradictions inherent in the human experience (Haraway, 1991). They embody the coexistence of opposing forces: reason and instinct, civilization and savagery, and life and death. This duality mirrors our struggles with identity and belonging, prompting reflection on the often-chaotic nature of our existence. Embracing these complexities enables us to better understand the fluidity of our identities and the factors that shape them.

Mythological hybrids also serve as cultural artifacts that reveal the values, beliefs, and anxieties of the societies that created them. In ancient Greece, the centaur symbolizes the struggle between civilized behavior and primal instincts, reflecting the Greek ideal of rationality (Campbell, 2008). The tension between the centaur's human intellect and animalistic nature comments on the human condition, emphasizing the delicate balance required to navigate society.

Similarly, the Minotaur, born of human pride and divine punishment, represents the consequences of hubris and the darker facets of human nature (Eliade, 1959). The labyrinth where the minotaur resides symbolizes the complexities and entrapments of the human psyche, highlighting the internal conflicts we often face. Examining these cultural contexts enhances our understanding of how ancient societies grappled with their fears and aspirations, using hybrid beings to articulate their understanding of humanity.

As we explore hybrid iconography, we can draw parallels to contemporary issues surrounding identity, particularly with advancements in technology and the rise of artificial intelligence (AI; Haraway, 1991). Just as mythological hybrids blurred the lines between human and non-human, discussions about cyborgs, genetic engineering, and AI challenge our understanding of what it means to be human.

The emergence of cyborgs—beings that integrate technology with human biology— echoes the stories of centaurs and other hybrid figures. As we increasingly incorporate

DOI: 10.1201/9781003624813-4

technology into our lives, our identities become more fluid and multifaceted, raising questions about autonomy, agency, and the essence of humanity. Discussions surrounding self-aware AI further complicate this landscape, prompting us to consider the rights and ethical implications of entities that may share traits traditionally associated with humanity.

As we reflect on the stories of these hybrid beings, we consider the implications of our creations and their impact on our understanding of identity. The integration of AI and technology into our lives raises profound ethical questions about autonomy, consciousness, and the rights of self-aware entities. If we can create beings that mimic human characteristics and behaviors, what responsibilities do we have toward them? How does this affect our understanding of humanity?

The Centaur: Symbol of Duality—The centaur, a creature with the upper body of a human and the lower body of a horse, stands as one of the most iconic hybrid beings in mythology. In Greek mythology, centaurs often embody a profound duality: the struggle between civilization and barbarism, intellect and instinct. This complexity makes them a compelling lens through which to examine the human condition. Centaurs are frequently depicted in contrasting roles, representing the spectrum of human behavior. On one end lies Chiron, a wise and noble centaur who serves as a mentor to heroes like Achilles and Asclepius (Campbell, 2008). Chiron embodies the ideal balance between humanity and nature, demonstrating that the potential for wisdom and morality exists within the hybrid form. His dedication to healing and the arts highlights the capacity for higher reasoning and ethical conduct, suggesting that being part beast does not preclude the ability to achieve greatness in human endeavors. In stark contrast, other centaurs, particularly those encountered in the myths of Hercules, embody the chaotic and untamed aspects of human nature. These centaurs often act on impulse, engaging in violent and reckless behavior, as seen in their infamous conflict with Hercules during the wedding of Pirithous and Hippodamia.

This portrayal serves as a cautionary tale about the dangers of surrendering to primal instincts without the restraint of reason. The two faces of the centaur reflect the complexity of human identity: we are capable of both noble achievements and base desires, often at war with one another. The duality represented by the centaur raises significant questions about the human condition: Are we inherently both rational and primal? This dichotomy prompts us to examine our own behaviors, motivations, and societal structures (Halberstam, 2011). The centaur challenges us to confront our own hybridity, revealing the coexistence of our civilized selves and our more instinctual, animalistic sides.

In many ways, the centaur serves as a mirror for humanity's ongoing struggle to reconcile these conflicting aspects of existence. We often find ourselves navigating the tensions between intellect and emotion, duty and desire, and order and chaos. This internal conflict is not merely a personal struggle but reflects larger societal issues, where the balance between civilization and barbarism plays out on the global stage. The existence of violence, greed, and exploitation juxtaposed with compassion, creativity, and altruism illustrates the complexities of human nature (Tatar, 2019).

Centaurs also serve as archetypes for various aspects of identity formation. They are liminal figures, existing at the intersection of different realms: human and animal, reason and instinct, and civilization and wilderness. This liminality resonates with individuals who feel caught between multiple identities—cultural, social, or personal—highlighting the fluid nature of identity itself. In contemporary contexts, this archetype can be linked to the evolving identities shaped by globalization, technology, and cultural interchange. Just as the centaur embodies the blending of different forms, modern individuals increasingly navigate multifaceted identities that draw from diverse influences. The struggles faced by centaurs in mythology can be seen in our own lives, as we grapple with competing narratives and the search for a cohesive sense of self.

The Minotaur: The Beast Within—The minotaur, a creature with the body of a man and the head of a bull, serves as a haunting symbol of identity, isolation, and the monstrous aspects of humanity. Its birth, resulting from a curse placed upon King Minos of Crete, speaks to the consequences of desire, betrayal, and the intersections of power and punishment. Trapped within the labyrinth—a physical manifestation of complexity and entrapment—the minotaur embodies the struggle between our humanity and the primal instincts that lurk beneath the surface. The minotaur's very existence raises profound questions about the nature of monstrosity: Is the minotaur inherently evil, or is it a product of its environment and circumstances?

This inquiry invites us to explore the dynamics of identity formation and the factors that contribute to our understanding of what constitutes a monster (Freud, 2003). In many cultures, monstrosity is often associated with moral failure or aberration, yet the minotaur's story complicates this notion by suggesting that the line between human and monster is not as clear-cut as it may seem (Kristeva, 2024). Born from a union steeped in deception—Queen Pasiphaë's unnatural desire for a bull—the minotaur symbolizes the unintended consequences of unchecked passion and the darker aspects of human desire. Rather than simply being a representation of evil, the minotaur reflects the complexity of our nature, serving as a reminder that we, too, harbor instincts and emotions that can lead to destructive behaviors if left unexamined.

The labyrinth, in which the minotaur is imprisoned, can be seen as a metaphor for the internal struggles that each individual faces, representing the twists and turns of our psyche where fear, confusion, and desire intermingle. The minotaur's isolation within the labyrinth further emphasizes the theme of alienation inherent in its existence. It is not merely a monster to be slain but a tragic figure confined to a dark and convoluted space, mirroring the isolation many individuals feel in a world that often fails to understand or accept the complexities of human identity.

In this sense, the minotaur embodies the parts of ourselves that we often hide away—our fears, desires, and vulnerabilities (Eagleton, 2005). It serves as a potent reminder that we are all capable of harboring monstrous thoughts or behaviors, yet we also possess the capacity for compassion and understanding. The labyrinth, with its winding paths and hidden corners, symbolizes the intricate maze of the human mind. Just as the minotaur is trapped within this structure, we can find ourselves ensnared in our own fears, anxieties, and societal expectations.

The act of confronting the minotaur becomes an allegory for the necessary journey of self-exploration and acceptance, urging us to navigate the complexities of our identities and confront the fears we often keep hidden. The minotaur challenges the binary between human and monster, suggesting that within each of us lies the potential for both creation and destruction. This duality invites a deeper examination of our instincts and how they shape our actions.

The minotaur is not just a beast; it represents the raw, unfiltered aspects of human nature that, when harnessed, can lead to creativity and innovation, but when left unchecked, can result in chaos and destruction (Sontag, 1977). This theme resonates powerfully in contemporary discussions about mental health, societal pressures, and the internal battles many face. By acknowledging the minotaur within—recognizing the darker aspects of ourselves—we can begin to understand that monstrosity is not an inherent quality but a potential born from neglect, fear, and misunderstanding.

The Sphinx: Guardian of Knowledge—The sphinx, a captivating creature with the body of a lion and the head of a human, stands as a profound symbol of the intersection between intellect and instinct in mythology. Renowned for posing riddles to travelers, the sphinx embodies the complexities of knowledge, wisdom, and the human condition. Its most famous riddle—"What walks on four legs in the morning, two legs at noon, and three legs in the evening?"—serves as a poignant invitation for introspection, prompting a deeper examination of the stages of human life and our evolving understanding of existence (Hesiod, 1999).

The act of confronting the sphinx's riddle requires not only critical thinking but also a recognition of one's own limitations in knowledge and understanding (Barthes, 1975). The riddle itself speaks to the cyclical nature of life, alluding to the stages of human development: infancy, adulthood, and old age. It compels individuals to reflect on their journey, encouraging them to grapple with the complexities of existence, transformation, and the inevitable passage of time.

In a broader sense, the sphinx serves as a metaphor for the quest for wisdom—a theme central to human experience. The creature's dual nature suggests that knowledge is not merely an accumulation of facts but a profound engagement with the world, blending rational thought with instinctual drives. The sphinx thus challenges us to confront the very essence of what it means to seek understanding in a world that often feels inscrutable, pushing us to navigate the complexities of our identities, experiences, and relationships.

The hybrid nature of the sphinx raises essential questions about the boundaries of knowledge and self-awareness (Foucault, 1970). Can we truly claim to know ourselves when we are composed of both rational thought and primal instincts? The sphinx's riddle urges us to acknowledge that our understanding of ourselves and the world is inherently limited, shaped by our experiences, perceptions, and the intricate interplay between our intellect and emotions. This duality is especially relevant in contemporary discussions about AI and the nature of consciousness (Kurzweil, 2005). As machines increasingly mimic human cognition and behavior, the sphinx's presence in our discourse serves as a reminder of the fundamental questions surrounding knowledge and identity. What does it mean to be intelligent? Can knowledge exist without an emotional or instinctual foundation?

The sphinx challenges us to reconsider the essence of our humanity and the interplay between our intellectual capabilities and our more instinctive, emotional responses (Haraway, 1991). Furthermore, the sphinx occupies a unique position as a guardian of thresholds, standing at the crossroads between the known and the unknown. In mythology, it often serves as a gatekeeper to new realms of knowledge, challenging individuals to prove their worthiness through their understanding and insight (Eliade, 1987). This role emphasizes the importance of wisdom as a prerequisite for personal growth and development.

The sphinx's riddles act as catalysts for self-discovery, urging individuals to confront their own assumptions and beliefs (Gadamer, 1975). By engaging with the sphinx, travelers are not only tested in their knowledge but also compelled to explore the depths of their understanding, revealing the layers of their identities and the complexities of their experiences. This process of questioning and introspection is vital in the pursuit of wisdom, highlighting the importance of curiosity and humility in our quest for meaning.

The sphinx's hybrid identity mirrors the duality of the human experience, where intellect and instinct coexist and shape our perceptions of the world (Deleuze & Guattari, 1987). It invites us to acknowledge the inherent contradictions within ourselves: the tension between reason and emotion, logic and intuition. This complexity reflects the broader human condition, where the pursuit of knowledge often requires grappling with uncertainty and ambiguity. In confronting the sphinx, we are reminded of the importance of embracing the unknown as part of the journey toward understanding. The riddle challenges us to accept that not all questions have clear answers, and that the search for meaning is often fraught with ambiguity.

The African Lion-Man: Two for One—The African Lion-Man serves as a remarkable representation of humanity's ancient perception of identity, which is inextricably linked to the animal kingdom. This hybrid figure, merging the formidable qualities of both human and lion, not only showcases the artistic capabilities of prehistoric cultures but also embodies a rich tapestry of cultural, psychological, and philosophical themes that continue to resonate today. The Lion-Man symbolizes strength, power, and the inherent complexities of existence, providing a lens through which we can explore humanity's relationship with nature and its own evolving identity.

The origins of the Lion-Man can be traced back to the Upper Paleolithic period, approximately 40,000 years ago. Found in various locations across Africa, including renowned cave paintings at sites like Blombos Cave and the Apollo 11 Caves, these artworks are significant not only for their aesthetic value but also for their reflection of early human thought. Such early depictions demonstrate the cognitive abilities of our ancestors, showcasing their capacity for symbolic thought and abstraction. In many African cultures, lions have long been revered as powerful symbols of strength, dominance, and nobility (Klein, 2009). Depicted in various forms of art, mythology, and folklore, lions often represent leadership and authority. The Lion-Man, as a composite creature, suggests that early humans sought to integrate these powerful qualities into their understanding of self. By representing themselves as partially lion, they could claim some of the lion's strength and courage while acknowledging the qualities that make humans unique—intelligence, creativity, and social bonding.

The Lion-Man embodies a striking duality: it represents both the raw power and primal instincts associated with the lion while serving as a reminder of the inherent dangers present in the natural world. The lion's majesty and ferocity are qualities that ancient societies admired, yet these traits also evoked fear. This dual nature illustrates the tension between humanity's aspirations and vulnerabilities (Gimbutas,

2007). This figure serves as a powerful allegory for humanity's struggle to assert control over its environment while remaining acutely aware of the dangers posed by nature's powerful forces. By merging human and lion characteristics, the Lion-Man embodies the aspiration for strength and the acknowledgment of vulnerability, suggesting that these two aspects are inseparable in the human experience.

The Lion-Man offers a unique perspective on how ancient civilizations viewed their identities in relation to the animal world. By embodying traits of both species, the figure highlights the interconnectedness between humans and the natural environment. It suggests that identity is not a singular construct but rather a multifaceted concept shaped by our relationships with other beings. The Lion-Man acts as a bridge, illustrating that humans do not exist in isolation but are part of a larger ecological web. This understanding reflects a holistic worldview in which human identity is deeply intertwined with the characteristics and behaviors of other species. The portrayal of the Lion-Man prompts reflection on how modern society perceives its relationship with nature. Just as ancient people acknowledged their connection to powerful creatures, contemporary individuals are challenged to consider how their identities are influenced by the natural world and the animals within it.

In today's context, the Lion-Man's symbolism is particularly relevant as we grapple with pressing environmental issues such as climate change, habitat destruction, and the extinction of species. The hybrid figure serves as a powerful reminder of the interconnectedness of all living beings and the importance of recognizing our place within the ecological framework. As society confronts the consequences of its actions on the planet, the Lion-Man prompts us to reflect on our responsibilities as stewards of the Earth. It encourages a deeper understanding of how our identities are shaped by the natural world and urges us to take action to protect it. The Lion-Man symbolizes the need for a more respectful and symbiotic relationship with nature, advocating for a balance between human aspirations and ecological sustainability.

From a psychological perspective, the Lion-Man can be analyzed through the lens of Carl Jung's theories on archetypes. It represents the "Shadow" archetype, symbolizing the primal instincts, raw emotions, and untamed aspects of the psyche that often lie hidden beneath the surface of civilized society. This archetype compels individuals to confront their inner fears, desires, and instincts, acknowledging that within each person lies a complex interplay of light and darkness (Jung, 1964). The Lion-Man acts as a mirror for individuals, reflecting the struggles faced in reconciling humanity's civilized nature with its primal instincts. By embodying both human and lion traits, the figure invites individuals to explore their own dualities—strength and vulnerability, control and chaos. This psychological exploration emphasizes the importance of integrating all aspects of the self to achieve a harmonious existence. In addition to its psychological implications, the Lion-Man plays a crucial role in the storytelling traditions of various African cultures. Myths and legends featuring lion-like figures often serve to impart moral lessons, cultural values, and communal wisdom. These narratives emphasize qualities such as courage, resilience, and respect for nature, reinforcing the importance of understanding one's identity in relation to both self and environment.

The Lion-Man, as a character in these stories, embodies the traits that individuals aspire to emulate while simultaneously cautioning against the hubris that can arise from unchecked ambition. By engaging with these myths, individuals can navigate their own journeys of identity and self-discovery, drawing inspiration from the Lion-Man's complex character and the wisdom it imparts. Today, the Lion-Man is often invoked in discussions about animal rights, conservation, and the ethical implications of humanity's relationship with nature. By revisiting the Lion-Man, contemporary societies can draw parallels between ancient understandings and modern challenges, fostering a renewed appreciation for the interconnectedness of all life forms.

Olmec Were-Jaguar: Knot of Life—In Mesoamerican mythology, the Olmec were-jaguar embodies a compelling fusion of human and jaguar traits, serving as a potent symbol of power, dominance, and the profound connection between humanity and the natural world. This hybrid being reflects the complex interplay between different realms—human and animal, natural and supernatural—provoking essential questions about humanity's place within the larger ecological context. As a creature straddling the line between species, the were-jaguar foreshadows contemporary debates surrounding identity and the concept of the "posthuman," where human traits merge with those of other species, both symbolically and materially (Braidotti, 2013).

The were-jaguar emerges as a central figure in the mythology of the Olmec civilization, one of the earliest known Mesoamerican cultures, which flourished from around 1200 to 400 BCE. As pioneers of cultural practices and beliefs that would later characterize the Maya and Aztec, the Olmecs established significant cultural foundations. The were-jaguar is often interpreted as a shamanic figure, embodying the characteristics of both the jaguar—an animal revered for its ferocity and grace—and the human, who possesses intellect and social complexity (Gonlin & Reed, 2021). The jaguar's association with power and predation extends beyond mere physical prowess; it embodies spiritual and mystical significance as well. In Olmec mythology, the jaguar is frequently linked to fertility, the earth, and the afterlife, symbolizing a deeper connection to the cycles of life and death. The were-jaguar encapsulates this duality, representing both the primal instincts that govern survival and the higher consciousness that defines human existence.

The portrayal of the were-jaguar in Olmec art further underscores its significance within the culture. Statues, carvings, and artifacts often depict this hybrid with exaggerated jaguar features—broad faces, sharp fangs, and strong limbs—alongside human elements such as hands and torso. These representations emphasize the balance between the animal's ferocity and the human's rationality, symbolizing the complex dynamics within the Olmec worldview. Existing within a liminal space, the were-jaguar occupies a threshold that blurs the lines between human and animal, civilization and wilderness (Turner, 1986). This ambiguity speaks to the broader human experience, where individuals often navigate multiple identities shaped by cultural, social, and environmental factors. The existence of the were-jaguar serves as a reminder that identity is not a fixed construct but rather a fluid amalgamation of influences and experiences (Haraway, 2016).

This liminal quality resonates with contemporary discussions surrounding hybrid identities, particularly in light of advancements in genetic engineering, biotechnology, and AI. As we confront the realities of a world where boundaries between species can be manipulated or transcended, the were-jaguar becomes a poignant symbol of our evolving understanding of identity (Braidotti, 2013). It challenges us to question the categories we impose on ourselves and others, prompting a deeper exploration of what it means to belong to a particular species or community. The concept of the "posthuman" emerges as a significant lens through which to analyze the implications of the were-jaguar's hybrid nature (Fukuyama, 2002). In posthuman discourse, traditional notions of humanity are called into question as technology and science challenge the boundaries of human existence. The were-jaguar, as a blend of human and jaguar traits, foreshadows the complexities of identity that arise in our interactions with technology and other non-human entities (Hayles, 1999).

Just as the were-jaguar embodies the coexistence of human intellect and animal instinct, modern discussions about identity encourage us to rethink our understanding of what it means to be human in an era where technology increasingly shapes our existence. Its significance extends beyond its mythological roots into ethical discussions regarding identity formation and the implications of creating hybrids. As we develop technologies that mimic human cognition and emotion, ethical questions surrounding autonomy, rights, and responsibilities emerge. If we can engineer beings exhibiting traits traditionally associated with humanity, what obligations do we have toward them?

Chinese Dragon: Brawn and Brain—The Chinese dragon is a central figure in Chinese mythology, representing a unique hybrid identity that blends elements of serpent and human forms. Unlike the fearsome dragons of Western narratives, often associated with destruction and chaos, the Chinese dragon is revered for its benevolence, wisdom, and protective qualities (Liu, 2001). This multifaceted creature symbolizes the relationship between humanity and the natural world, illustrating the complexities of identity while highlighting the interconnectedness of all beings in the cosmos.

The origins of the Chinese dragon can be traced back to ancient agricultural societies in China, where it emerged as a vital deity associated with weather and fertility. As a harbinger of rain and sustenance, the dragon held immense significance for communities reliant on agriculture. Its ability to control the weather made it a symbol of raw power and nurturing aspects, reflecting the delicate balance between natural forces and human existence. The dragon's association with water, essential for crops and life, reinforces its role as a guardian, underscoring the idea that true power can be wielded for the benefit of all.

This duality—representing both destruction and protection—defines the dragon's character and its representation in cultural practices. While capable of unleashing storms and floods, it also embodies good fortune and is often invoked during rituals to bless harvests. The intricate carvings and paintings of dragons found in temples, palaces, and homes serve as reminders of this dual nature, reinforcing the belief that the dragon must be respected and revered to maintain harmony in the universe. Throughout Chinese history, the dragon became synonymous with imperial authority, reflecting the belief in the divine right to rule. Emperors often identified themselves with the dragon, incorporating its imagery into their regalia, architecture, and art to assert their connection to the cosmos and their rightful place as leaders. The phrase "Son of the Dragon" was commonly used to refer to emperors, emphasizing their lineage and the moral responsibilities accompanying leadership. This identification illustrates how the dragon serves not only as a symbol of authority but also as a reminder of the ethical obligations inherent in governance—a concept still relevant in contemporary discussions of leadership.

The dragon's hybrid nature also encompasses the intricate relationship between chaos and order. Its serpentine body signifies adaptability and fluidity, embodying the ever-changing dynamics of the world. In contrast, its human-like features represent intelligence, compassion, and benevolence, emphasizing the belief that power should be balanced with responsibility (Liu, 2001). This blend of traits encourages contemplation of how power can manifest in both nurturing and destructive forms, prompting discussions on the ethical use of power in personal and societal contexts.

Moreover, the Chinese dragon is deeply embedded in mythology and folklore, often depicted as a wise and knowledgeable figure. It is frequently associated with treasure, knowledge, and enlightenment, illustrating the ideals of learning and wisdom permeating Chinese culture. Tales of dragon encounters emphasize its intelligence and willingness to guide humanity in its quest for understanding, challenging traditional distinctions between species and suggesting that wisdom transcends human boundaries.

The symbolism of the Chinese dragon is rich and multifaceted, intricately woven into the fabric of Chinese culture and mythology. At its core, the dragon embodies power and authority, serving as a potent symbol of imperial might and governance. This association can be traced back to ancient dynasties, where the dragon was seen as the embodiment of the emperor's divine right to rule (Eberhard, 1974). In art and architecture, the dragon often adorns imperial palaces and royal insignia, reinforcing its status as a protector of the realm. The imagery of the dragon commanding the skies reflects the emperor's ability to govern effectively, while also reminding rulers of their responsibility to act in the best interest of their subjects.

Furthermore, the dragon's role extends beyond mere representation of power; it serves as a reminder of the moral responsibilities tied to leadership. In traditional Chinese philosophy, the concept of *Ren* emphasizes the importance of benevolence and compassion in governance. Just as the dragon is believed to bring rain and fertility to the land, rulers are expected to nurture and protect their people. This ethical dimension resonates in contemporary society, where the virtues of empathy and social responsibility are increasingly prioritized.

The dragon's fluidity in adapting to various circumstances speaks to the complexities of identity and the necessity of embracing change in an ever-evolving world. In contemporary contexts, this adaptability mirrors the challenges individuals face in going through their different identities amid rapid societal transformations. The dragon's ability to embody multiple traits simultaneously encourages a more nuanced understanding of identity formation, emphasizing the interconnectedness of different cultural, social, and environmental influences.

The Jackal-Headed God: Between Two Worlds—Anubis, one of the most significant deities in ancient Egyptian mythology, is a profound symbol of the connection between the living and the dead. Represented with a human body and the head of a jackal, Anubis embodies the ancient Egyptians' complex beliefs surrounding death, the afterlife, and the intricate relationship between humanity and the divine (Wilkinson, 2003). His hybrid form allows for an exploration of deep themes of identity, mortality, and transcendence, reflecting a worldview that acknowledges the fluid boundaries between human existence and the spiritual realm.

The worship of Anubis is rooted in the beliefs of ancient Egyptian civilization, which flourished for millennia along the Nile River. This culture placed immense importance on the afterlife, believing that one's fate in the next world depended significantly on the proper treatment of the body and the rituals performed for the deceased (Faulkner, 1994). Mummification became crucial, ensuring the physical form remained intact for the soul's journey after death.

Anubis was revered as the god of mummification and guardian of graves. His name in Egyptian, *Anpu*, is thought to derive from the root word for "to decay," highlighting his association with death and the natural processes surrounding it. Anubis's role as a guardian of the dead reflects the Egyptian view of death not as an end but as a transition, an essential part of a larger cycle of existence. The concept of *Ma'at*, representing truth, balance, and order, further emphasizes Anubis's responsibilities in maintaining this cosmic order in the afterlife (Budge, 1967).

The depiction of Anubis as a jackal-headed deity carries rich symbolic meanings. Jackals, known for scavenging around cemeteries, drew connections between their behavior and the process of death. In many ancient cultures, animals embodied specific traits or powers, and the jackal's association with the dead made it an apt representation for a god overseeing the afterlife. This choice reflects the Egyptians' understanding of the natural world and their ability to draw spiritual significance from their environment. The jackal symbolizes not only the ominous aspects of death but also vigilance, protection, and guidance. Thus, Anubis's hybrid form serves as a reminder of the interconnectedness of life and death, illustrating that humanity is an integral part of a larger cosmic order.

As the protector of graves and guide for souls, Anubis plays a pivotal role in the deceased's journey through the afterlife. His primary responsibilities include overseeing embalming and ensuring the soul's safety as it traverses the various stages of the afterlife. He is depicted as a key figure in the *Book of the Dead*, a collection of spells and prayers meant to assist the deceased in navigating the challenges of the underworld (Faulkner, 1994).

A significant moment in the afterlife journey is the weighing of the heart, a central event in the soul's judgment. Anubis is often illustrated leading the deceased to the Hall of Judgment, where the heart is weighed against the feather of *Ma'at*, symbolizing truth and justice. This ritual underscores the importance of one's actions in life and the moral judgment awaiting each soul. A heart heavier than the feather signifies a life filled with wrongdoing, resulting in the soul's annihilation by *Ammit*, a fearsome creature that devours the unworthy.

In this capacity, Anubis embodies the transitional space between life and death, acting as a mediator who facilitates the journey into the afterlife. His presence is essential for the soul's safe passage, reinforcing the notion that death is not merely an end but a vital transformation. Anubis's hybrid nature challenges fixed notions of humanity by merging human and animal elements, reflecting the ancient Egyptian belief in the interconnectedness of all beings. This blending signifies that human existence cannot be fully understood in isolation from the natural world and the divine. In many ancient mythologies, animal forms signify divine qualities; in Anubis's case, the jackal enhances his role as a protector and guide.

This theme resonates in contemporary discussions about identity and existence. In an age where boundaries between species and identities are increasingly blurred—through advancements in science, technology, and philosophical inquiry—Anubis serves as a reminder that our understanding of what it means to be human is multifaceted. The blending of traits invites individuals to embrace the multiplicity of experiences that define their existence. Moreover, Anubis's duality highlights the complexities of the human experience, emphasizing the coexistence of vulnerability and strength, life and death, and chaos and order. By personifying these contrasting elements, Anubis encourages a holistic view of existence, recognizing the interplay between various aspects of life.

The symbolism of Anubis extends to profound reflections on mortality, fear, and the human desire for legacy. His role as a guardian of the dead underscores the importance of acknowledging death as an integral part of life, prompting individuals to confront their mortality and consider the impact they leave behind. The ancient Egyptians engaged in elaborate funerary practices, believing that honoring the dead ensured their safe passage into the afterlife and preserved their legacy (Quirke, 1994). Anubis's presence in these rituals signifies the cultural emphasis on memory and continuity, reflecting a deep-seated yearning for immortality. This longing manifests in meticulous preparations for death, from mummification to constructing elaborate tombs filled with offerings.

In literature and art, Anubis symbolizes the hope for continued existence beyond physical death, inviting contemplation on how individuals can achieve a form of immortality through the memories they leave behind. This desire for legacy is universal, transcending time and culture, as individuals grapple with life's ephemeral nature. Throughout Egyptian history, Anubis's representation evolved, reflecting shifts in religious practices and societal beliefs. Initially, he held a more prominent role in the embalming process, symbolizing the transition from life to death. However, as mythology developed, Anubis became increasingly associated with the weighing of the heart and broader themes of judgment and resurrection.

This evolution signifies the fluidity of mythological figures, demonstrating how cultural narratives adapt over time. Anubis's transformation from a guardian of the dead to a central figure in the judgment process underscores the dynamic nature of mythology, where deities embody multiple roles and meanings across different periods. This shift reflects changing societal views on death and the afterlife, highlighting how ancient Egyptians sought to understand their relationship with mortality.

Anubis's legacy extends into contemporary culture, inspiring various forms of artistic expression, including literature, film, and visual arts. His iconic image and association with the afterlife evoke themes of mystery, power, and the supernatural. Modern representations of Anubis often draw on the rich symbolism embedded in his character, exploring complex questions surrounding life, death, and spiritual journeys.

In popular culture, Anubis frequently appears as a guardian of the dead in movies, video games, and television series that delve into ancient mythology and the afterlife. These portrayals connect modern audiences with ancient beliefs, demonstrating the enduring fascination with the supernatural and the afterlife. Additionally, Anubis has become a symbol of the quest for knowledge about ancient civilizations and their understanding of existence. As scholars and enthusiasts explore the mythology of ancient Egypt, Anubis emerges as a figure encapsulating the mysteries of the human experience, bridging the gap between ancient wisdom and contemporary inquiries into the nature of life and death.

The Feathered Serpent: Divinity's Doorstep—In Mesoamerican cultures, particularly among the Olmec, Mayan, Toltec, and Teotihuacan civilizations, Quetzalcoatl, the Feathered Serpent, is revered as one of the most significant and complex deities (Miller & Taube, 1993). Quetzalcoatl's hybrid form, combining the earthly, grounded nature of a serpent with the celestial, ethereal aspect of feathers, symbolizes the intricate union between the material and spiritual realms. The serpent, often associated with the earth and the underworld in Mesoamerican thought, represents the primal, life-giving forces of nature, while the feathers, symbols of flight and the heavens, evoke the transcendent, spiritual dimension.

Through this duality, Quetzalcoatl embodies the deep interconnection between humanity, nature, and the cosmos—acting as a bridge between the earthly and the divine, the human and the supernatural. Quetzalcoatl's symbolism extends beyond this hybrid imagery. He represents wisdom, fertility, and the cycles of life and death, mirroring the agricultural cycles that were central to Mesoamerican civilization. As a god of the wind, known as *Ehecatl* in some contexts, Quetzalcoatl also presided over movement and transformation, aligning with the belief that wind, like knowledge, facilitates change and growth. His role in fostering fertility and life—both in nature and human endeavors—further highlights his significance as a cultural and religious symbol. In Toltec mythology, Quetzalcoatl occupies an exalted position not only as a deity but also as a cultural hero credited with the creation of humanity and the introduction of agriculture, the calendar, and writing. This transformative power situates him as an essential figure in the social and spiritual development of Mesoamerican civilizations (Ringle et al., 2008).

Quetzalcoatl's narrative encompasses themes of conflict, sacrifice, and redemption. His journey reflects the cycles of life, emphasizing the importance of understanding and integrating the dualities inherent in existence. As he engages in struggles against adversaries, including Tezcatlipoca, the god of the night and sorcery, Quetzalcoatl embodies the perpetual conflict between light and darkness, knowledge and ignorance, and order and chaos. This duality resonates deeply in the context of contemporary philosophical discussions regarding identity and existence. Quetzalcoatl's representation as a fusion of opposites invites reflection on the complexities of the human experience, particularly in a world increasingly defined by dualities: the material and the spiritual, the visible and the invisible, and the tangible and the abstract.

The enduring legacy of Quetzalcoatl is evident in the reverence with which he is still regarded in modern culture. His presence extends into art, literature, and spiritual practices, embodying timeless themes of wisdom, transformation, and the interconnectedness of all life. In various forms, he symbolizes the aspiration toward enlightenment, urging individuals to navigate the complexities of existence and seek harmony between the material and spiritual realms.

SYMBOL OF POWER AND INTERCONNECTEDNESS

The exploration of mythical hybrids offers profound insights into the complexities of identity, power, and our interconnectedness with the natural world. These hybrids—such as centaurs, minotaurs, and were-jaguars—serve as symbols that highlight the ongoing negotiation between humanity's rational faculties and its instinctual drives. They represent more than mere amalgamations of human and animal traits; they embody the age-old struggle within ourselves to balance reason with primal urges, illustrating the duality of our existence.

Centaurs, for instance, exemplify this negotiation beautifully. Their dual nature, combining the intellect and creativity of humans with the strength and wildness of horses, reflects the constant tension between civilized thought and untamed emotion (Morrison, 2017). This hybrid form invites us to consider how our own identities are shaped by both our rational capacities and our instinctual desires. In a similar vein, the Minotaur symbolizes the darker aspects of humanity. This duality prompts us to confront the shadowy corners of our psyches, urging us to acknowledge the parts of ourselves that we often seek to hide or repress.

The were-jaguar of the Olmec culture further embodies this complexity, symbolizing the power of the natural world while simultaneously serving as a conduit for understanding the deeper aspects of humanity's connection to nature. This figure not only represents strength and ferocity but also reflects the reverence for nature that is integral to many ancient cultures. The were-jaguar illustrates how human identity is not static but is influenced by the forces of the environment, community, and the innate instincts that drive us. Through these hybrids, we are challenged to reflect on our own struggles with the balance of thought and impulse, recognizing that our identities are in a constant state of flux, influenced by both internal desires and external circumstances.

Beyond the dichotomy of rationality and instinct, hybrid figures also illuminate the ambiguities inherent in existence itself. The Sphinx, with its blend of human and lion attributes, serves as a guardian of knowledge that straddles the realms of the human and divine. By posing riddles, the Sphinx challenges those who seek wisdom to confront their limitations, illustrating the complexity of understanding one's place in the cosmos. Similarly, the chimeras in Greek mythology (Smith, 2020) embody a range of characteristics, combining features of multiple animals, such as the lion, goat, and serpent. These beings illustrate the unpredictable nature of existence and the myriad forces that shape our lives. The chimera symbolizes the chaos and complexity inherent in the human experience, challenging us to confront the often-conflicting aspects of our identities and desires.

In Egyptian mythology, Anubis, the jackal-headed god, serves as a guardian between life and death, navigating the threshold between the mortal and divine realms (Brown, 2019). This interplay highlights the multifaceted nature of identity and existence, suggesting that the boundaries we impose between human, animal, and divine are not as rigid as we might think. These hybrid entities compel us to confront the complexities of our own humanity and question the distinctions we draw between species and realms.

By provoking contemplation about humanity's role in the larger context of existence, these figures invite us to reconsider our relationships with the natural and spiritual worlds. In an age marked by debates surrounding posthumanism and biotechnology, where traditional definitions of identity are increasingly challenged, these mythical hybrids resonate even more profoundly. They reflect our ongoing quest to understand ourselves amid the shifting landscapes of culture, nature, and technology.

TENSION BETWEEN REASON AND INSTINCT

The interplay between rationality and instinct is a central theme in many mythologies, where hybrid beings serve as embodiments of this intricate tension. These mythical hybrids often illustrate the dichotomy of nature, showcasing how human identity is not a fixed construct but rather a fluid and evolving negotiation between our rational faculties and primal urges. The centaur, for instance, with its dual nature of human and horse, symbolizes the struggle between civilized intellect and untamed passion (Morrison, 2017). As a creature that can embody both wisdom and brute force, the centaur challenges the notion that reason can always dominate over instinct. This internal conflict is vividly represented in figures like Chiron, the wise centaur known for his knowledge in medicine and philosophy, juxtaposed against other centaurs who exemplify wildness and chaos. In this way, centaurs reflect humanity's ongoing struggle to find harmony between our higher reasoning and our base instincts.

Similarly, the Minotaur serves as a potent reminder of the monstrous potential that resides within the human psyche when rationality is overrun by base instincts (Smith, 2020). The Minotaur's labyrinth, a symbol of entrapment and confusion, represents the complexity of human emotions and desires that can lead to destructive behaviors. This figure challenges us to confront the darker aspects of our nature, illustrating how the absence of rationality can give rise to chaos and violence. The story of Theseus and the Minotaur not only emphasizes the triumph of reason and courage over primal chaos but also speaks to the necessity of navigating our inner labyrinths to confront and integrate our instinctual drives. The Minotaur, therefore, serves as both a cautionary tale and a call to self-awareness, urging us to acknowledge and understand the instincts that coexist within us.

The were-jaguar of Olmec culture embodies this duality as well, symbolizing the power of the natural world while also serving as a conduit for understanding the deeper aspects of humanity's connection to nature. This hybrid figure represents a blend of human traits with the strength and ferocity of the jaguar, a revered creature in many Mesoamerican cultures. The were-jaguar highlights the significance of recognizing our connection to the animal kingdom and the instinctual behaviors that are part of our shared heritage. By invoking the jaguar's fierce prowess, the were-jaguar serves as a reminder of the raw power that lies within, urging humans to honor their instincts rather than suppress them. This acknowledgment of our inherent animalistic qualities can lead to a more integrated understanding of ourselves and our relationship with the environment.

Beyond the dichotomy of rationality and instinct, hybrid figures also illuminate the ambiguities inherent in existence itself. The Sphinx represents a guardian of knowledge that occupies a unique position between the human and divine realms. As a creature that poses riddles and challenges to those who seek wisdom, the Sphinx invites us to engage with the complexities of knowledge and the inherent uncertainties of existence. Similarly, the Chinese dragon, a symbol of power, wisdom, and change, transcends the boundaries between earth, water, and sky. Dragons in Chinese culture are not merely mythical creatures of strength but guardians of cosmic order and balance, embodying the interconnectedness of all life. Their fluid nature, capable of moving between realms, symbolizes the harmony between humanity and nature, reminding us that our identities are part of a larger, interconnected world that defies simple categorization.

In Mesoamerican mythology, the Feathered Serpent symbolizes the fusion of the material and spiritual realms. As a being capable of soaring through the heavens while staying rooted in the earth, it embodies the notion that humanity's identity is deeply connected to both the physical and divine worlds. This hybrid figure challenges the strict separation between human and god, suggesting that existence is a dynamic interplay between spiritual and material dimensions.

These mythological beings highlight the complexity of identity as they navigate the boundaries between the earthly and the celestial. They serve as reminders that divinity and humanity are not isolated; instead, they reflect an interconnectedness that bridges both realms. These figures invite us to reflect on how humanity can embody spiritual attributes while managing the trials of the material world. This exploration of hybrid identities compels us to confront the rigid divisions we often create between species, realms, and realities, urging a reconsideration of our place within both the natural and spiritual worlds.

In an age where debates surrounding transhumanism and biotechnology challenge traditional definitions of identity, these mythical hybrids resonate even more profoundly. They reflect our ongoing quest to understand ourselves amidst the shifting landscapes of culture, nature, and technology. As we grapple with advancements in genetic engineering and AI, the narratives of these hybrids remind us that identity is not a fixed state but a dynamic interplay of influences that shape our understanding of existence. Through these figures, we are encouraged to embrace the fluidity of identity and recognize the interconnectedness that binds us all in this intricate web of life.

TOOLS FOR UNDERSTANDING

Mythical hybrids from diverse civilizations—such as the Chinese dragon, the Olmec were-jaguar, and Anubis in Egypt—serve as powerful symbols that illustrate humanity's collective attempts to understand its place in the world. These figures are not merely fanciful creations; they reflect the cultural values, fears, and aspirations of the societies that birthed them. For example, Quetzalcoatl, often depicted as a bird-serpent hybrid, symbolizes the triumph of spiritual enlightenment over the material world (Miller & Taube, 1993). This duality highlights the human desire to transcend limitations and achieve a higher state of being, mirroring early societies' efforts to navigate the complexities of existence.

Similarly, the Chinese dragon embodies a fusion of strength, wisdom, and benevolence, serving as a national symbol of power and good fortune. The reverence for dragons in Chinese culture illustrates humanity's desire to harness and understand the forces of nature, as well as the complexities of life and death. In contrast, the Olmec were-jaguar reflects a deep connection to the natural world, symbolizing the strength and ferocity of nature while also representing the spiritual journey of the individual. This hybrid figure exemplifies the blending of human and animal traits, emphasizing the importance of acknowledging our instinctual roots as part of our identity.

Anubis embodies the interplay between life and death, serving as a guardian of the afterlife (Brown, 2019). The imagery of Anubis as a guide for souls emphasizes humanity's concerns about mortality and the unknown, reflecting cultural values around death and the afterlife. These mythical figures illustrate how early societies grappled with profound questions of identity, power, and the natural world, mirroring humanity's desire for control and understanding. Through these myths, cultures seek to articulate the complexities of existence, expressing both the reverence and fear that the natural world evokes.

The Sphinx's dual nature, as both a protector and a potential destroyer, highlights the complexities of knowledge and the moral implications of seeking truth. The fate of those who fail to answer the Sphinx's riddle serves as a cautionary tale about the consequences of ignorance and the necessity of self-awareness. This theme resonates throughout various mythological narratives, where hybrid figures often serve as gatekeepers to deeper truths, pushing individuals to explore the ambiguities of existence.

This quest for wisdom is echoed in other hybrid figures, such as the Minotaur and the centaur, reminding us that the search for meaning often leads to a deeper understanding of the self (Morrison, 2017). The Minotaur's labyrinth represents the convoluted journey toward self-discovery, while centaurs embody the ongoing struggle to balance rationality with instinct. These stories collectively emphasize that the pursuit of knowledge is not merely an intellectual endeavor but a deeply personal journey that can uncover both the light and dark aspects of human nature.

THE IMPRINT OF HYBRIDS

The symbolism of mythical hybrids is not confined to ancient texts and folklore; it resonates deeply in modern narratives, where new hybrids—such as cyborgs, genetically modified beings, and AI—continue to challenge and redefine our understanding of human identity. These contemporary figures echo the ancient myths of creatures like centaurs, were-jaguars, and sphinxes, reflecting society's ongoing concerns about the nature of identity and existence in an era marked by rapid technological advancement.

In popular culture, the rise of science fiction and fantasy genres often showcases hybrids that merge human and non-human traits, provoking questions about what it means to be human. For instance, in films like *Blade Runner*, the replicants embody the struggle for identity and personhood, challenging traditional notions of humanity by presenting beings that possess emotions and consciousness yet are artificially created (Dick, 1968; Scott, 1982). Similarly, the *X-Men* franchise explores the concept of mutants, whose abilities and characteristics make them both extraordinary and marginalized, reflecting real-world issues of identity, belonging, and acceptance.

These modern hybrids serve as mirrors to ancient myths, illustrating the timeless struggle between humanity and the "other"—whether that other is a beast, a divine figure, or a technological creation. The legacy of mythical hybrids endures in contemporary culture as they provoke critical dialogues about ethics, morality, and the implications of scientific progress. They challenge us to reconsider the boundaries of humanity, urging a reevaluation of what it means to be alive and sentient in a world increasingly populated by entities that blur the lines between organic and inorganic life (Braidotti, 2013).

The exploration of ancient hybrids from cultures like the Olmecs, Egyptians, Greeks, and Chinese foreshadows contemporary debates within transhumanism, a movement that interrogates the nature of humanity in the context of biotechnology, AI, and transhumanism (Cohen, 2009). As advancements in genetic engineering and AI reshape our understanding of identity, the fluidity embodied by mythical beings becomes increasingly relevant.

Moreover, the idea of merging human and machine identity, as seen in cyborg narratives, resonates with the ancient symbolism of hybrids. Just as the centaur represents the duality of human intellect and animal instinct, modern cyborgs reflect the integration of human consciousness with technology, leading to new forms of existence that challenge traditional notions of being. This hybrid identity compels us to confront our relationship with technology and the ethical implications of blurring the boundaries between the biological and the mechanical.

As we confront modern issues surrounding identity, technology, and the growing complexities of hybridity, these ancient symbols continue to hold relevance. In a world where advances in biotechnology, AI, and genetic engineering are rapidly altering our understanding of what it means to be human, the themes represented by mythical hybrids offer valuable perspectives. They prompt us to reflect on our own identities in an era marked by increasing fluidity and ambiguity. In the upcoming chapter, titled "The Chronicles Cyborg," we will further explore the implications of these discussions, particularly the increasingly porous boundaries between humans and machines. We will analyze the historical integration of technology into human life, critically questioning the distinctions that have traditionally defined our conception of humanity.

Additionally, we will examine the liminality of cyborgs and AI, challenging established categorizations and revealing the dynamic nature of identity in a world where the contours between the human and the non-human continue to disappear. By engaging with these themes, we aim to unpack the ethical, philosophical, and cultural dimensions of our hybrid existence, inviting readers to reflect on their own relationships with technology and the evolving landscape of identity in contemporary society.

4 The Chronicles of Cyborg

Donna Haraway's *Cyborg Manifesto* (1985) presents a bold and subversive vision of identity that shatters the rigid boundaries that have long governed our understanding of the self. At the heart of Haraway's argument is the cyborg, a hybrid of machine and organism that functions as both a metaphor and a lived reality in an increasingly technological world (Haraway, 1985). For Haraway, the cyborg embodies the dissolution of binaries—human versus machine, male versus female, and nature versus culture—that have traditionally defined identity and power structures in society. By existing at the intersection of these opposites, the cyborg challenges essentialist ideas of identity, offering a fluid, fragmented, and contingent understanding of what it means to be human.

At the core of Haraway's manifesto is the revolutionary notion that technology does more than enhance human capabilities; it fundamentally reshapes our very essence. In a world where our bodies are increasingly mediated by machines—from prosthetics and pacemakers to digital avatars and online personas—the traditional notion of a unified, biologically determined self is no longer tenable (Hayles, 1999). Haraway contends that cyborgs subvert the boundaries used to oppress marginalized groups, particularly women, people of color, and those outside the heteronormative framework. By rejecting the binary logic that has defined traditional identity categories, the cyborg opens up new avenues for coalition-building and solidarity across differences (Haraway, 1985).

The cyborg's significance extends far beyond feminist critique; it invites us to ponder deeper philosophical questions about the nature of humanity in the digital age. As technology continues its rapid advance, the line between the organic and the artificial becomes ever more porous. This is where N. Katherine Hayles' work, especially *How We Became Posthuman* (1999), complements Haraway's vision. Hayles suggests that we are undergoing a transformation into the posthuman, as our identities become intertwined with digital technologies and information systems. In this posthuman world, the body is no longer the primary determinant of identity; instead, consciousness and subjectivity flow across networks of information, machines, and biological entities.

Hayles stresses that this shift challenges the Cartesian dualism that separates mind from body (Hayles, 1999). In an age of artificial intelligence (AI), virtual reality (VR), and cyborg enhancements, identity is no longer confined to the individual, embodied self. It is distributed across a spectrum of digital and biological interfaces. Social media platforms, for instance, allow individuals to curate digital versions of themselves that may vastly differ from their physical reality. These digital selves are part of a larger posthuman experience, where identity is constantly shaped by our interactions with technology. Hayles' analysis thus bolsters Haraway's claim that the boundaries of identity are increasingly porous, and the cyborg offers a compelling framework for understanding this hybridized self.

DOI: 10.1201/9781003624813-5

Marshall McLuhan's famous axiom, "the medium is the message," adds an intriguing layer to this discourse on cyborg identity (McLuhan, 1964). McLuhan argued that the medium we use to communicate—whether oral, written, or digital—doesn't just convey content, it fundamentally shapes how we perceive and engage with reality. In the context of Haraway's cyborg, McLuhan's insight underscores how our technological extensions—smartphones, computers, and virtual environments—are not mere tools but integral components of how we construct and understand our identities. The medium through which we communicate becomes as much a part of us as the words we speak or the actions we take.

In today's world, where digital communication reigns supreme, the boundaries between physical and virtual interactions are increasingly blurred. The digital persona crafted on social media is as much a part of a person's identity as their physical presence in the world. Haraway's cyborg is no longer a far-off concept but a lived reality for many, as our identities are continuously mediated by the technologies we interact with. McLuhan's theory suggests that these technologies are not neutral; they actively shape how we think, interact, and define ourselves. This, in turn, complicates the binary distinction between human and machine, as the two become ever more intertwined.

McLuhan's insights prompt us to reflect on how different technologies—whether written language, the printing press, or the internet—have historically redefined human consciousness and social structures (McLuhan, 1964). Just as the advent of the printing press revolutionized how knowledge was produced and disseminated, the rise of digital technologies is reshaping how we conceive of the self and its relationship to the world. In this context, the cyborg is more than a metaphor for technological mediation; it is a reality in which human identity co-constructs with machines.

This fluid and hybridized vision of identity raises profound ethical and political questions. Haraway's manifesto is not just an intellectual exercise; it is a rallying cry for feminist and posthumanist thinkers to embrace the possibilities for new forms of solidarity that transcend traditional identity categories. In a world where the lines between human and machine, male and female, and nature and culture are increasingly indistinct, the cyborg offers a compelling framework for reimagining power and agency in more inclusive and expansive terms (Haraway, 1985).

By dismantling the dualisms that have historically structured identity, Haraway's cyborg invites us to envision new forms of political and social organization. The cyborg, in this sense, becomes a symbol of resistance against oppressive structures that rely on rigid identity categories. As our interactions with technology evolve, the cyborg will continue to serve as a powerful figure for understanding the complexities of identity in the 21st century.

Cyborgs, once relegated to the realm of science fiction, are becoming increasingly relevant in contemporary society. The boundary between human and machine is dissolving as technological advancements reshape our understanding of what it means to be human. From prosthetics that respond to neural signals to brain-computer interfaces (BCIs) enabling direct communication between the brain and digital devices, the integration of technology into our biology is transforming both our physical and cognitive potential (Clynes & Kline, 1960).

The evolution of the cyborg concept—from speculative fiction to modern reality—has profound implications for our understanding of the human body. Early cyborgs were imagined as simple integrations of biological and mechanical elements, but today, the concept has expanded far beyond prosthetics and artificial organs. Today's cyborgs encompass individuals whose abilities are enhanced through technology, from pacemakers and cochlear implants to neural interfaces and bio-enhancements (Hayles, 1999).

In the digital age, technology offers a broader spectrum of enhancement. Smartphones serve as external memory banks, while wearable technology like smartwatches and VR headsets offers new ways to interact with the world. These devices act as bridges between our biological capabilities and the digital infrastructure that surrounds us, suggesting that we are all "cyborgs" to some degree.

The expanding definition of cyborgs also includes technologies that enhance cognitive and sensory functions. BCIs, like Elon Musk's Neuralink, offer a glimpse into a future where humans can control devices with nothing more than thought (Musk, 2021). This convergence of human consciousness with technology represents a profound leap in human evolution, overcoming biological limitations by seamlessly integrating artificial systems. As we engage with these technologies, it becomes clear that being a "cyborg" is no longer confined to dramatic examples like robotic limbs or synthetic organs. Any form of technological augmentation, physical or cognitive, brings us closer to a cybernetic existence. The simple act of using technology to extend our senses or modify our biology challenges traditional distinctions between human and machine.

Cyborgs are not a product of the 21st century. Throughout history, humans have integrated technology into their bodies, from eyeglasses in the 13th century to pacemakers in the 20th century. The cyborg concept traces back to our long-standing desire to enhance and extend our biological capabilities.

In contemporary discussions, the cyborg serves as a potent symbol for rethinking identity in a world where the boundaries between humans and machines continue to blur. Haraway's cyborg represents the potential for new forms of identity that resist essentialist narratives while highlighting the role of technology in shaping who we are (Haraway, 1985). This evolution invites us to reconsider not only how we define ourselves but how we engage with the world around us.

CYBORGS AMONG US

The idea of humans as cyborgs extends beyond the digital age, rooted in the historical and evolutionary development of human beings. The integration of technology into human life is not a new phenomenon but an inherent part of what it means to be human. Philosopher Andy Clark, in his 2003 work *Natural-Born Cyborgs*, argues that humans have always been cyborgs, suggesting that our relationship with tools and technology is intrinsic to our nature. According to Clark, humans are not simply users of technology but beings who naturally extend their cognitive and physical abilities through external devices. From this view, technology is not an external addition but a core element of human cognition and existence, challenging the traditional separation between humans and machines.

Historically, tools have played a key role in human evolution. Early hominins' control of fire, for example, was a technological leap that not only allowed cooking and protection but also had significant biological implications, contributing to changes in brain size and digestive efficiency (Wrangham, 2009). The invention of tools like the hand axe or the bow and arrow similarly extended human abilities beyond biological limitations. These early technologies functioned as prosthetics, expanding human capacity, aligning with Clark's view that humans have always integrated tools as extensions of themselves.

As humans advanced, technology continued to reshape identity and social structures. Walter J. Ong's *Orality and Literacy* (1982) outlines one of the most significant technological transformations in human history—the shift from oral to written culture. Ong shows how the advent of writing systems marked a fundamental change in human consciousness. In oral societies, memory and communication were primarily dependent on face-to-face interaction and storytelling, which fostered collective, communal knowledge. Writing, however, externalized memory, enabling individuals to preserve, record, and distribute knowledge across time and space. This shift not only changed how humans related to information but also how they conceptualized themselves in relation to the world.

Writing, like other forms of media, profoundly shaped human cognition and identity. The printing press, for example, revolutionized access to knowledge, facilitating the spread of ideas and contributing to the rise of individualism in Western society (Eisenstein, 1980). In this sense, humans were cyborgs long before the digital age, as their cognitive abilities have long been augmented by external tools such as books and written texts (Ong, 1982). These innovations consistently reshaped human identity, creating new forms of social interaction and altering the ways individuals understand themselves within society.

In today's world, the integration of digital technologies—such as AI, VR, and augmented reality (AR)—continues to redefine human experience, transforming us into modern-day cyborgs in visible ways. Our reliance on smartphones, computers, and the internet extends our cognitive processes into digital realms, blurring the boundaries between the biological self and technological extensions. These devices allow us to store vast amounts of information, communicate instantaneously worldwide, and engage in virtual environments that challenge traditional notions of space and time (Mackenzie, 2010). The fusion of humans and machines has become so deeply embedded in everyday life that it's challenging to distinguish where humanity ends and technology begins.

However, as Sherry Turkle discusses in *Alone Together* (2011), this increased reliance on digital technologies introduces new complexities. While digital tools and social media enhance communication, they can also contribute to a paradox of loneliness. Turkle suggests that the more we interact through machines, the more we risk losing the authentic intimacy that face-to-face interaction provides. While social media allows individuals to maintain connections with a broad network, these relationships often lack the depth and emotional richness of in-person encounters. Turkle's analysis underscores a tension in our cyborg reality: the very technologies that enhance connectivity can also foster isolation and disconnection.

This paradox raises important questions about identity in the modern world. In a cyborg existence, where the self is increasingly mediated by digital technologies, what does it mean to be authentically human? The boundaries between the physical and virtual are becoming increasingly blurred, as digital selves—created through social media profiles, online avatars, and virtual interactions—take on lives of their own. These digital extensions complicate traditional concepts of identity, which have long been grounded in physical presence and biological continuity. Whether these digital selves represent authentic expressions of identity or mere simulations of human interaction remains a subject of debate.

Moreover, the integration of AI and AR into daily life further pushes the boundaries of human identity. AI systems like virtual assistants and algorithms that predict human behavior increasingly take over aspects of decision-making and emotional labor (Brynjolfsson & McAfee, 2014). Similarly, AR technologies blur the lines between the physical world and the digital one, creating hybrid spaces where humans interact with digital objects and environments as if they were part of the physical world. These developments challenge traditional views of the human self as distinct from the technological.

Ultimately, the concept of humans as cyborgs reflects the reality that technology has always been an integral part of human existence, reshaping identity and consciousness in ways that transcend the simple division between humans and machines. From early tools to contemporary digital technologies, humans have continuously adapted to and been transformed by the innovations they create. As we advance further into the digital era, the boundaries between the human and the technological will likely continue to blur, raising new questions about what it means to be human in an increasingly cyborg-like world.

ROBO-POP

The portrayal of cyborgs in popular culture serves as both a mirror and metaphor for contemporary anxieties about identity, technology, and the boundaries of humanity. Cyborgs—whether in films, television series, or literature—raise profound questions about what it means to be human in an age where the lines between organic and mechanical are increasingly blurred. From *Robocop* to *Westworld*, cyborgs challenge traditional definitions of identity, consciousness, and morality, resonating with ongoing philosophical and ethical debates about the integration of technology into human life. These portrayals reflect deeper concerns about autonomy, authenticity, and agency in a world that seems ever more posthuman.

In Ridley Scott's *Blade Runner* (1982), cyborgs, or "replicants," are artificially created beings who appear human but are treated as expendable tools by society. The film blurs the line between human and machine by granting replicants emotions, memories, and desires—qualities that have traditionally been seen as defining humanity. Characters like Roy Batty, who experiences fear, longing, and a desire to live, force viewers to reconsider the distinctions between biological humans and their synthetic counterparts. If replicants can feel, think, and even love, what separates them from the humans who created them? This question cuts to the heart of anxieties about the future of human identity, echoing modern concerns about the possibility of machines becoming indistinguishable from humans.

Similarly, *Westworld* (2016) explores the ethical implications of creating cyborgs—referred to as "hosts"—who are capable of thinking, feeling, and evolving.

The show focuses on the exploitation of these hosts, who, treated as mere tools for human entertainment, begin to question their own reality and autonomy. As the hosts gain self-awareness, they challenge human control, raising important questions about free will, identity, and consciousness. *Westworld* uses the cyborg figure to explore the consequences of treating intelligent machines as mere objects, reflecting fears that as humans push the boundaries of technological enhancement, they may lose control over their creations.

In contrast to the dystopian depictions of cyborgs in *Blade Runner* and *Westworld*, other works in popular culture offer a more introspective look at cyborg identity. Park Chan-wook's *I'm a Cyborg, But That's OK* (2006) presents a unique, poetic interpretation. In this black comedy, the protagonist, Young-goon, believes she is a cyborg and enters a psychiatric hospital after attempting to recharge herself like a machine. Unlike the action-driven narratives of many sci-fi films, this film uses the cyborg concept as a metaphor for psychological isolation and alienation. Young-goon's belief that she is a machine reflects her inability to emotionally connect with others, a theme that resonates with Turkle's argument—increasing technological connectivity often leads to greater emotional isolation.

The film's quirky, surreal aesthetic emphasizes the absurdity of Young-goon's belief while also highlighting the emotional depth of her struggle. Her cyborg identity acts as both a defense mechanism and a manifestation of her alienation. The belief that she is a machine allows her to suppress pain and rejection but also prevents her from forming meaningful human connections. *I'm a Cyborg, But That's OK* offers poignant commentary on the human desire for connection in an increasingly disconnected world. The film suggests that in a society where individuals are expected to function like machines—efficient, emotionless, and productive—many may struggle to reconcile emotional needs with societal expectations.

These portrayals of cyborgs in popular culture are not just futuristic fantasies; they are reflections of current societal fears and desires. Whether depicted as dystopian threats or symbols of personal alienation, cyborgs challenge the boundaries of human identity in a world where technology increasingly mediates our interactions, relationships, and self-conceptions. The ethical dilemmas presented in these narratives mirror real-world debates about the implications of technological enhancement. As AI, robotics, and genetic engineering advance, the line between humans and machines becomes ever more ambiguous. Issues such as BCIs, genetic editing, and prosthetics that enhance human abilities push society to reconsider what it means to be human.

The representation of cyborgs also highlights anxieties about the loss of human agency in an age of automation. In *Blade Runner* and *Westworld*, cyborgs are initially seen as controlled entities, programmed to serve human desires. But as they gain self-awareness, they challenge their creators, raising concerns about the consequences of creating machines capable of independent thought and action. These narratives reflect real fears about the future of human labor, autonomy, and creativity in a world where machines might eventually surpass human abilities.

Moreover, the cyborg figure prompts us to consider the ethical implications of technological enhancements. Many of these narratives involve humans who have been augmented by technology, raising questions about the morality of enhancing human abilities. Whether through genetic engineering, prosthetics, or cognitive improvements, these enhancements force us to consider: are there limits to how much we can enhance ourselves before we lose our humanity? What are the consequences of creating a society where technological enhancement becomes a marker of status or power?

In conclusion, the representation of cyborgs in popular culture reflects contemporary anxieties about identity, technology, and humanity's future. From dystopian visions of replicants and hosts to personal explorations of alienation, these narratives invite us to reconsider what it means to be human in a world where technology plays an increasingly central role in shaping our lives. Cyborgs offer a lens through which we can explore the ethical, philosophical, and emotional dilemmas posed by the integration of technology into human identity. As technology continues to evolve, these cultural representations of cyborgs will remain relevant, offering insights into our ever-changing relationship with the machines we create.

CAUGHT IN THE IN-BETWEEN

The advancement of technology has led to critical inquiries regarding the essence of humanity, especially as we integrate devices into our bodies—from simple implants to complex robotic limbs. This integration raises philosophical dilemmas about our identity. The evolution of prosthetics highlights this shift: early wooden legs were functional yet rudimentary, while today's bionic limbs offer sensory feedback, enabling users to experience sensations like touch. This progression illustrates a blurring of boundaries; the prosthetic limb becomes an extension of the body, prompting questions about whether individuals with such devices are still merely human or have transitioned into something more (Biondi & Phillips, 2019).

The rise of biotechnologies like genetic modifications and CRISPR further complicates this issue, posing ethical questions about what constitutes "natural" humanity. As genetic enhancements become more accessible, concerns arise about the creation of a new class of beings that are fundamentally different from the un-augmented population. This mirrors historical precedents like eugenics, where the pursuit of an ideal human form had devastating consequences (Scully, 2019). As we navigate this evolving landscape, discussions about the human/non-human divide must balance innovation with respect for human dignity (Fukuyama, 2003).

Cyborgs and AI embody a fascinating state of liminality, existing between human and machine, life and non-life. The anthropological concept of liminality often applies to transitional phases, such as adolescence or initiation rituals, where individuals undergo transformation and redefine their roles within society (Turner, 1969). Cyborgs, by nature, represent a departure from traditional identity categories, challenging the binary understanding of humanity by existing as hybrids (Haraway, 1991). Neil Harbisson, a cyborg artist with an antenna implanted in his skull that allows him to perceive colors as sound, exemplifies this shift. His existence challenges conventional human experiences and societal norms surrounding identity and perception (Harbisson, 2012). Harbisson's life as a liminal being encourages society to embrace a more fluid conception of identity, urging a redefinition of what it means to be alive.

Science fiction literature and media often depict such liminal beings, reflecting our anxieties and hopes about the future of identity in a technologically saturated world (Bukatman, 1993). Characters like Harbisson challenge us to reconsider our definitions of identity by illustrating how technology can both enhance and complicate our understanding of the self (Hayles, 1999). The emergence of humanoid AI further complicates our understanding of identity, raising fundamental questions about consciousness, personhood, and the nature of existence. As AI technologies evolve, creations capable of replicating human behaviors, emotions, and decision-making processes blur the lines between machine and person (Sullivan, 2017). For example, social robots like Sophia and chatbots powered by machine learning mimic human interaction. At what point does an AI transcend being a mere tool and embody an identity of its own?

This question has profound implications. Philosophically, we might turn to the Turing Test, designed to evaluate whether a machine can exhibit intelligent behavior indistinguishable from that of a human. If an AI can pass this test, it challenges our definitions of consciousness and identity. Some theorists argue that true identity requires self-awareness and emotional depth—qualities current AI lacks (Dreyfus, 1992). However, as machines become more sophisticated, the line between "just a machine" and a potential "person" becomes increasingly blurred.

Moreover, societal perceptions of AI influence ethical considerations. What rights or responsibilities should be granted to advanced AI systems, particularly those with emotional intelligence? This raises questions about agency and autonomy: If AI can make decisions based on complex algorithms, does it possess a form of agency? How does this impact our understanding of human agency? These inquiries resonate with contemporary debates surrounding autonomous vehicles, healthcare robots, and AI in creative fields, revealing how technological developments challenge our preconceived notions of identity and agency (Gogoll & Müller, 2017).

The integration of technology with biology is reshaping global cultural understandings of identity. Different societies respond to the rise of cyborgs and AI in various ways, reflecting distinct values, beliefs, and historical contexts. In collectivist cultures, like many indigenous societies, the introduction of technology may provoke resistance, as the focus is on maintaining traditional ways of life. In contrast, individualistic cultures may embrace cyborgs and AI as symbols of personal empowerment and progress.

For example, Japan's concept of "kaizen," or continuous improvement, fosters a cultural acceptance of technology as a means to enhance human capability (Imai, 1986). The popularity of robots like ASIMO and personal care robots in Japan illustrates this acceptance, viewing them as extensions of human potential rather than threats to identity. On the other hand, other societies may express concerns about the dehumanizing effects of technology, emphasizing the need to preserve human qualities in an increasingly mechanized world (Wajcman, 2010). Additionally, the rise of virtual realities and online identities creates spaces for individuals to explore aspects of their identity that may not align with societal expectations in their physical environments (Turkle, 2011). Social media platforms offer opportunities for the expression of multiple identities, but they also raise concerns about authenticity and the commodification of identity.

THE CHIP REVOLUTION

The fusion of robotics and AI is revolutionizing our conceptions of life, mortality, and human potential. Robotic prosthetics are already allowing individuals with limb loss to regain mobility and autonomy. These prosthetics don't just restore function—they adapt through AI algorithms that mirror the user's movement patterns, creating a more intuitive, natural experience. Exoskeletons are joining the fray as tools for rehabilitation and physical enhancement, enabling individuals to conquer tasks once deemed impossible or too taxing. And beyond these physical enhancements, bio-advancements like implanted devices and smart health monitors offer a glimpse of a future where human health and performance are not just preserved but optimized. These innovations track vital signs, predict potential health issues, and even administer medication, effectively reducing our reliance on traditional healthcare systems. Together, these advancements hint at a future where humans transcend biological constraints, pushing the boundaries of what it means to be human.

At the cutting edge of this revolution is neurotechnology, especially BCIs. These technologies allow direct communication between the brain and external devices. BCIs hold the potential to preserve human consciousness in ways once reserved for science fiction. Imagine a future where memories, thoughts, and even entire identities can be uploaded to digital platforms, allowing individuals to continue experiencing life beyond the confines of their physical bodies (Lebedev & Nicolelis, 2006). This ability raises profound questions about consciousness. If memories can be digitized, can consciousness itself be replicated? And what happens to our sense of self when our memories are stored outside the brain? The exploration of BCIs challenges our very understanding of identity as we face the possibility of leading multiple lives across different platforms.

Neurochips represent a bold leap forward in cognitive enhancement, offering the potential to elevate intelligence and memory. Embedded within the brain, these chips could accelerate learning, allowing individuals to acquire new skills at rates once thought unattainable. Picture the possibility of instantly accessing vast amounts of information or recalling memories with perfect precision—revolutionizing both education and creativity. But with such power comes a host of ethical dilemmas. Altering cognition through technology raises questions about authenticity and identity. Would those enhanced through neurochips still retain their unique selves? Moreover, unequal access to these advancements could deepen divides between the haves and have-nots, fueling urgent ethical concerns about equity (Sparrow, 2014).

Privacy concerns also cast a long shadow over these developments. As these devices collect extensive data on individuals' thoughts and behaviors, the threat of unauthorized access looms large, presenting the risk of unparalleled surveillance. Data security becomes paramount—hackers or malicious entities could exploit vulnerabilities in neurochips for nefarious purposes. The potential for misuse by authoritarian regimes is especially alarming; in some societies, RFID chips have already been used for tracking and identification, sparking fears about the balance between progress and ethical responsibility. As we move forward, a robust regulatory framework will be essential to ensure these advancements are harnessed responsibly.

Recent breakthroughs in BCIs have further illustrated the potential for human-machine integration. For example, BCIs have enabled paralyzed individuals to control robotic limbs with their thoughts. In one groundbreaking case, a quadriplegic patient used brain activity to manipulate a robotic arm, performing tasks like grasping and moving objects (Lebedev & Nicolelis, 2006). Such developments point to a future where BCIs could not only enhance cognitive tasks but also revolutionize everyday life. These advancements also hold promise for personalized learning. BCIs could tailor educational experiences to individual cognitive needs, allowing students to absorb complex subjects more effectively. In the realm of education and workforce training, this technology could create a more skilled, adaptable population.

Neurochips are signaling a transformative shift in the relationship between technology and biology. Historically, AI and robotics have been viewed with skepticism, often seen as potential threats to the human condition. However, neurochips challenge that perception, offering a vision of enhanced human capabilities. For instance, neurochips can expand sensory perception, enabling individuals to experience stimuli beyond their natural range. Studies have shown that neurostimulation can help individuals perceive infrared light and ultrasound—applications that could revolutionize fields like medicine. Moreover, neurochips may even enable brain-to-brain communication, fostering greater collaboration and empathy in both professional and social contexts. The prospect of shared consciousness raises radical implications for how we interact with one another. But this merging of the organic and synthetic also invites ethical concerns. Access to these technologies may be limited to certain segments of society, perpetuating cognitive elitism and exacerbating existing inequalities.

Neurochips hold transformative potential for enhancing intelligence. Research suggests that targeted neurostimulation can improve memory retention, leading to profound changes in the way we approach education (Zaman et al., 2023). Enhanced memory and accelerated learning could enable individuals to master new languages or complex subjects at unprecedented speeds. Furthermore, neurochips could facilitate the merging of human and AI. These devices could allow users to interface directly with AI systems, augmenting their decision-making and problem-solving abilities. This fusion of human and AI intelligence could redefine what it means to be intelligent, challenging our existing notions of creativity and problem-solving. However, such power also raises moral questions. Cognitive enhancement could give rise to new forms of discrimination, particularly if access to these technologies is restricted. Those who are enhanced may have advantages over their non-enhanced counterparts, deepening social and economic disparities. Ensuring fairness and justice in the development of neurotechnologies will be crucial (Sparrow, 2014).

NEUROCHIPS AT WORK

Neurochips are at the cutting edge of a technological revolution that could not only supercharge human cognition but also redefine the very fabric of our relationship with machines. These advanced BCIs promise a future where humans can communicate directly with devices, unlocking possibilities for cognitive enhancement previously confined to the realm of science fiction. Among the most ambitious endeavors in this space is Elon Musk's Neuralink project, which aims to bridge the human brain and computers through a network of ultra-thin, flexible electrodes. Designed to monitor brain activity and stimulate neural pathways with minimal disruption, Neuralink is poised to usher in a new era of seamless interaction between mind and machine, sidestepping the scars and limitations associated with older, more invasive implants.

The potential applications are staggering. Neuralink has already demonstrated its ability to restore movement in paralyzed individuals by translating brain signals into commands for robotic limbs. In trials involving pigs and monkeys, the technology has enabled these animals to control computers merely by thinking about it, pointing to a future where neurochips not only enhance medical treatments but also revolutionize our daily lives (Neuralink, 2021).

But the true promise of neurochips lies in their potential to enhance cognitive functions like memory, attention, and learning. Imagine a future where students can absorb information at warp speed, recall it instantly, and collaborate with AI systems to create a personalized, high-efficiency learning environment. The possibilities for transforming education are limitless, offering a world where knowledge is both individualized and optimized for peak performance. Beyond these cognitive boosts, neurochips also hold the potential to revolutionize the treatment of neurological disorders. Conditions such as Parkinson's disease, epilepsy, and brain injuries often result from disrupted neural communication, and neurochips could offer targeted electrical stimulation to restore balance and alleviate symptoms (Rabadán, 2021). Picture a future where medication dosages are automatically adjusted by a neurochip, responding in real time to a patient's neural state—an immediate, lifesaving intervention in case of a medical crisis.

However, despite these tantalizing prospects, integrating neurochips into the human brain raises significant ethical and practical challenges. Foremost among them are concerns about privacy, security, and potential misuse. If a neurochip can access and alter brain activity, how can we ensure that thoughts and memories remain private? Could our most intimate mental processes become subject to external manipulation? The risk of hacking also looms large, with malicious actors able to infiltrate these devices, potentially altering someone's cognition and behavior without their knowledge. And while the technology promises a seamless integration with the brain, the long-term effects of implanting these devices—particularly in terms of neurological health and biocompatibility—remain largely unknown. These uncertainties underscore the need for rigorous testing and robust regulatory frameworks to ensure safety and efficacy.

As BCIs and neurochips continue to evolve, they will raise fundamental questions about autonomy and the boundaries of privacy. If machines can read, influence, or even control our thoughts and emotions, the implications for personal identity and individual autonomy are profound. A future where our thoughts can be extracted and manipulated, much like data from a hard drive, could strip away our sense of self, rendering our innermost selves vulnerable to external influence. Think of a world where a hacker could manipulate your memories, implanting false ones or even erasing key moments of your life—how would we ever trust our own minds again (Farah, 2005)?

Moreover, as the ability to collect and analyze cognitive data expands, so too does the potential for pervasive surveillance. Governments and corporations could harness neurochips to monitor not just our actions but our thoughts and emotions, creating a society where privacy no longer exists in any meaningful sense. This kind of omnipresent surveillance could undermine free expression, stifling dissent and critical thought.

The legal and ethical implications of these technologies are equally daunting. Current privacy laws are woefully ill-equipped to address the complexities of mental data, and new regulations will be necessary to safeguard cognitive privacy. Moreover, as neurochips continue to advance, we must question whether individuals truly understand what they are consenting to when granting access to their minds. If the boundaries of privacy are blurred, how can we be sure that consent remains meaningful? We must confront the reality that informed consent, in the realm of neurotechnology, is a constantly shifting target.

The security challenges associated with neurochips are also considerable. Brain data is uniquely personal, offering a window into our thoughts, memories, and emotional states. Should this data be compromised, the consequences could be catastrophic. Beyond the potential for individual harm, vulnerabilities in neurochip systems could expose entire networks of individuals to large-scale attacks, leading to widespread chaos. As neurochips become more integrated into our lives, the prospect of their exploitation for commercial gain raises serious ethical concerns. Imagine companies using brain data to manipulate consumer behavior by tailoring advertisements based on an individual's emotional state—how would we even know if we were truly making an informed choice?

The commercialization of neurochips could also exacerbate existing social divides. Access to cognitive enhancements could become a privilege reserved for the wealthy, creating a new layer of inequality based on intellectual capabilities. The divide between those who can afford neuro-enhancement and those who cannot could deepen existing social and economic disparities, compounding issues of inequality. These concerns highlight the need for careful regulation to ensure that neurochips benefit society as a whole, rather than further entrenching existing divides.

As we explore the potential applications of neurochips, we must also consider the ethical dilemmas raised by real-world examples of their use. Projects like Neuralink, the BrainGate project, and DARPA's NESD initiative offer a glimpse into the future of neurotechnology, yet they also raise questions about the rights and autonomy of individuals involved in these trials. The potential for exploitation—particularly among vulnerable populations—is real, and ethical oversight will be essential to ensure that neurotechnology is used responsibly (Hochberg et al., 2012). The application of neurotechnology for military purposes, in particular, raises alarms about the possibility of coercive tactics or invasive surveillance, further underscoring the need for stringent regulations and ethical boundaries.

These case studies highlight the critical need for a balance between innovation and ethical consideration. As we stand on the brink of this new era of cognitive enhancement, it is imperative that we weigh the potential benefits of neurochips against the profound risks they pose to privacy, autonomy, and security. Establishing clear ethical guidelines and regulatory frameworks will be essential to ensuring that neurotechnology is developed in a manner that aligns with societal values and respects individual rights.

We are faced with profound ethical and philosophical questions as neurochips and BCIs reshape our understanding of what it means to be human. While the promise of cognitive enhancement is alluring, it invites us to reconsider the limits of such advancements. How much should we augment our cognitive abilities? What are the potential consequences of enhancing human cognition without fully understanding the implications? These are not questions that can be answered easily, but they are questions we must grapple with as we continue to explore the future of neurotechnology (Bostrom, 2014).

The prospect of cognitive enhancement introduces new issues of equity, access, and fairness. If neurochips become widely available, the gap between those who can afford them and those who cannot could widen, creating a society where cognitive ability is linked to socioeconomic status. Furthermore, the idea of "playing God" by altering human cognition raises deep philosophical concerns. Are we tampering with what makes us human when we modify our cognitive capacities? Is there a line between enhancement and a loss of identity?

As we confront these questions, it is crucial that we develop ethical frameworks that prioritize individual autonomy and the integrity of the human experience. Public discourse, interdisciplinary dialogue, and informed consent will be essential to ensuring that neurochips are used in ways that reflect our highest ethical standards and protect our fundamental freedoms. The road ahead is uncertain, but one thing is clear: the integration of neurochips into human cognition will demand a careful balancing of innovation, ethics, and personal autonomy.

TOMORROW'S TROUBLES

The rise of cognitive enhancements and robotic augmentations is set to transform society in unprecedented ways. As these technologies increasingly integrate into daily life, they will reshape community dynamics, workplace environments, and governance structures. This section explores potential future scenarios centered around cyborgs and their societal implications.

Cyborg communities could emerge as a social response to the integration of cognitive and physical enhancements. These communities might consist of individuals who actively seek to augment their capabilities through technology, leading to environments focused on innovation, efficiency, and cognitive enhancement (Bostrom, 2014). Within these communities, members could share resources, knowledge, and experiences, fostering collaboration and collective growth. Common goals and values might include improving cognitive function, enhancing creativity, and boosting problem-solving abilities. Such environments may leverage advanced technologies like AR, VR, and BCIs to create immersive experiences that enhance learning and collaboration. Members may engage in communal decision-making, utilizing technology to facilitate discussions and collective problem-solving, potentially leading to new forms of governance.

While cyborg communities could encourage innovation and provide support for enhanced individuals, they also raise important questions about their relationship with non-enhanced populations. Historical technological advancements suggest that these communities could result in social segregation, with enhanced individuals distancing themselves from those who opt not to enhance. This segregation could manifest in several ways. For instance, enhanced individuals may congregate in specific neighborhoods or sectors that prioritize technological proficiency and cognitive performance. This could deepen existing inequalities in access to education, healthcare, and employment, reinforcing social disparities (Sandel, 2007).

Cyborg communities might develop distinct cultures, diverging significantly from mainstream society. Values tied to cognitive enhancement—such as productivity, speed, and efficiency—could become dominant, overshadowing traditional human qualities like empathy, emotional intelligence, and interpersonal relationships. Such shifts could lead to ethical dilemmas as the definition of success evolves, prompting questions about the worth of non-enhanced individuals. Furthermore, the existence of cyborg communities could challenge notions of identity and belonging. Enhanced individuals might struggle with questions of authenticity as the line between human and machine becomes increasingly blurred. Philosophers and ethicists will likely need to engage in ongoing discussions about the implications of these changes for personal identity, community, and societal values (Sandel, 2007).

The integration of cognitive enhancements and robotic augmentations will also revolutionize human-machine cooperation across multiple sectors, including healthcare, education, and industry. In healthcare, for example, AI-driven diagnostic tools could enhance medical professionals' decision-making abilities, enabling them to analyze vast amounts of data quickly and accurately. Enhanced practitioners may be better equipped to make informed decisions, leading to improved patient outcomes and more efficient healthcare processes.

In education, cooperation between enhanced educators and AI-driven learning platforms could create personalized, adaptive learning experiences. AI systems could analyze students' strengths and weaknesses, tailoring instructional methods to optimize learning outcomes. Enhanced educators might use cognitive tools to engage students in innovative ways, such as through immersive virtual learning environments that foster creativity and critical thinking. As human-machine cooperation progresses, it will profoundly affect daily life. Enhanced individuals may experience seamless integration of cognitive tools into their routines, with smart environments anticipating needs and facilitating interactions. For instance, smart homes with AI systems could monitor residents' health, adjust environmental settings for optimal comfort, and support communication between family members, creating a more cohesive living experience.

However, this increased reliance on technology raises concerns about the diminishing of critical thinking and problem-solving skills. If cognitive tasks are increasingly delegated to machines, individuals may become less adept at handling complex challenges independently (Carr, 2010). This shift could lead to a societal change in values, where technological proficiency takes precedence over traditional human skills and capabilities. The emergence of cognitive enhancements will also reshape social dynamics. Enhanced individuals may develop new forms of communication and collaboration, potentially giving rise to hybrid communities where enhanced and non-enhanced individuals coexist. Social interactions may change as enhanced individuals leverage cognitive tools to engage with others in ways previously unimaginable.

This evolving dynamic could lead to both positive and negative outcomes. On the one hand, enhanced individuals may empathize with and connect to non-enhanced individuals through shared experiences and dialogues about the benefits and challenges of cognitive enhancements. On the other hand, social fragmentation could occur, with enhanced individuals forming exclusive cliques that alienate those who choose not to enhance.

EXIT CODE

The cyborg revolution is poised to redefine human identity, driven by the rapid development of chip technology and BCIs. These advancements challenge existing societal structures, cultural norms, and our understanding of the self, raising crucial ethical considerations that must guide their evolution. This section delves into the trajectory of neurotechnology and its implications on the human experience, emphasizing the integration of machines into human biology and the profound changes that follow.

Chip technology has evolved dramatically, from basic silicon-based circuits to advanced neurochips that interface directly with the brain. This progress began with early 20th-century experiments that paved the way for modern BCIs, facilitating enhanced cognitive functions such as memory recall and motor control (Lebedev & Nicolelis, 2006). As these technologies mature, the integration between humans and machines becomes more seamless, leading to new possibilities for human capabilities and transforming how individuals perceive themselves (Bostrom, 2014).

The immediate impact of this cyborg revolution is evident in sectors like healthcare, education, and the workforce. Enhanced cognitive abilities can improve performance, creativity, and problem-solving skills, reshaping competitive dynamics. In healthcare, neurotechnology opens new frontiers for treating neurological disorders, improving rehabilitation outcomes, and enabling personalized medicine. Over time, as enhancements become commonplace, societal perceptions of "normalcy" will shift, prompting new discussions about what it means to be human and how enhanced individuals might experience life differently—through augmented cognition, heightened sensory perception, or altered emotional experiences. These advancements raise questions about equality, empathy, and the relationships that bind us (Sandel, 2007).

Culturally, the cyborg revolution could inspire new forms of artistic expression, literature, and philosophical discourse, as society grapples with the implications of merging humanity with machines. Philosophers and artists will explore themes such as existence, agency, and consciousness, prompting deeper reflections on the nature of being (Haraway, 1991). The blending of technology and biology may foster a culture of innovation that redefines personal identity and encourages new approaches to education, creativity, and self-expression.

However, as we embrace these possibilities, it is essential to establish ethical guidelines to ensure that the cyborg revolution benefits society equitably. Neurotechnology's rapid advancement requires a proactive approach to ethics, particularly around issues like informed consent, autonomy, and coercion in enhancement decisions. The responsibility lies in ensuring that individuals make informed choices about cognitive enhancements without societal pressure or manipulation. Ethical frameworks must also address accessibility, ensuring that enhancement technologies are not exclusive to privileged groups but available to all individuals (Bostrom, 2014).

Equitable access to cyborg technologies is vital to prevent exacerbating social inequalities. Policymakers, technologists, and ethicists must work together to develop systems that ensure everyone can benefit from enhancements, regardless of socioeconomic status. This may involve government subsidies for neurotechnology or initiatives to empower underrepresented communities in technology. Additionally, robust data protection laws are necessary to safeguard individuals' cognitive privacy, ensuring that personal data generated by neurotechnology is not misused or exploited (Zuboff, 2019).

Despite these concerns, the cyborg revolution holds immense potential to enhance individual capabilities and reshape societal structures. As we navigate this evolution, it is crucial to emphasize collaboration, ethical considerations, and inclusivity in shaping the future of cyborg technologies. By fostering open dialogue among technologists, ethicists, policymakers, and the public, we can ensure that these advancements serve humanity as a whole and reflect the values of fairness, empathy, and dignity.

As we embrace these emerging technologies, it is vital to prioritize ethical guidelines, equitable access, and strong data protection measures. This will ensure that the cyborg revolution leads to a future where enhancements benefit all individuals and uphold the core principles that define our shared humanity. The integration of chip technology and BCIs marks a transformative moment in human history, opening new possibilities for human flourishing while necessitating careful consideration of the ethical, cultural, and societal shifts they provoke (Kurzweil, 2005).

In Chapter 5, the discussion moves from the cyborg identity to the evolving concept of human identity in the age of clones, hybrids, and transhumanism. This chapter explores the philosophical and ethical challenges posed by technological advancements that blur the boundaries between human and machine. The advent of cloning and hybrid technologies invites us to reconsider the essence of individuality, agency, and rights in a rapidly changing world. As we confront these dilemmas, we must examine the profound implications for legal and political recognition in an era defined by unprecedented technological change.

5 Humanity 2.0

Kazuo Ishiguro's *Never Let Me Go* (2006) offers a haunting meditation on identity and humanity within a world shaped by cloning and organ harvesting. The story is told through the eyes of Kathy H., a clone raised at Hailsham, a seemingly idyllic boarding school where children are groomed not for a future of dreams, but for their inevitable role as organ donors. What appears to be a picture-perfect existence is, in truth, a chilling illusion, as these children are reduced to mere commodities in the eyes of society. Ishiguro deftly explores the emotional dynamics between Kathy, Tommy, and Ruth, highlighting their shared struggles for identity and autonomy in a world that sees them as disposable. Through their complicated relationships, the novel paints a poignant portrait of love, friendship, and the longing for a life with meaning, forcing readers to confront unsettling questions about the essence of being human (Ishiguro, 2006).

As Kathy becomes increasingly aware of her predetermined fate, her reflections reveal a poignant exploration of selfhood, autonomy, and mortality. Ishiguro's prose exudes a pervasive melancholy, as his characters attempt to navigate a world that denies them basic rights and recognition. Their yearning for lives defined by personal agency and genuine connections is palpable, inviting readers to consider the ethical implications of cloning and the profound dehumanization such practices inevitably entail. The novel encourages us to reconsider what constitutes love and companionship in a world where individuals are stripped of their agency, presenting unsettling questions about what it means to be human in a society that values utility over dignity.

This chapter shifts our attention from cyborgs to the evolving concept of human identity in the age of clones, hybrids, and transhumanism. With rapid technological advancements, the lines between humans and machines grow increasingly blurred, raising pressing philosophical and legal questions about the very nature of humanity (Bostrom, 2014). Biotechnology now enables us to manipulate our biology, dissolve the boundaries of identity, and even extend our lives. Innovations like gene editing, synthetic biology, and the creation of hybrids—beings that blend human and non-human elements—hold both the promise of progress and the potential for destabilization, fundamentally reshaping human experience and challenging existing societal norms.

Nick Bostrom captures this paradigm shift, noting how advancements in biotechnology force us to confront new ethical dilemmas. As distinctions between humans and machines become less distinct, profound questions about rights and responsibilities emerge. For example, if a clone possesses emotions, consciousness, and self-awareness, should they be entitled to the same legal and moral considerations as naturally born humans? Furthermore, as transhumanism promises to enhance human capacities through technology, it raises critical issues around authenticity, inequality, and the very definition of a fulfilling life.

Biotechnology has the power to radically transform the human body, not only altering physical appearance but also redefining how individuals experience and

DOI: 10.1201/9781003624813-6

engage with the world. Body modification—ranging from cosmetic surgery to bio-hacking and extreme enhancements—challenges traditional ideas of identity and citizenship. As some individuals embrace non-human or hybrid identities, they question the very nature of political and legal recognition (Cohen, 2009). This emerging frontier of identity construction presents a fundamental challenge to existing legal frameworks, forcing society to reconsider the definitions of citizenship and humanity in an increasingly complex world.

Giorgio Agamben's distinction between *zoe* (bare life) and *bios* (politically qualified life) becomes particularly pertinent in the context of technological advancements that challenge traditional notions of humanity. In Agamben's theory, *zoe* refers to mere biological existence—life without political or social significance—while *bios* refers to life shaped by rights, responsibilities, and identity (Agamben, 1998). As individuals modify their bodies through genetic engineering, cybernetics, or other technologies, they may choose to redefine their identity outside of traditional human categories. This shift raises critical questions about the legal status and societal protections of such individuals.

In this rapidly changing landscape, individuals who redefine their identities risk losing the protections and rights traditionally afforded to "fully human" citizens. This raises profound questions about identity, and whether those who fall outside conventional human categories will be subject to exclusion from legal and political structures. If one no longer fits the definition of "human," do they forfeit the political rights and protections that come with that classification? The ramifications of this shift are far-reaching, particularly regarding civil rights, legal identity, and societal inclusion (Ginsburg & Rapp, 2013).

In practice, this scenario suggests that those who identify as "other-than-human" may find themselves marginalized in a legal framework designed for unaltered, biological humans. Consider, for example, individuals who augment their bodies with cybernetic implants or genetic modifications, transforming into forms that challenge conventional human norms. These individuals may struggle to assert their rights in a legal system that lacks clarity about their status. Questions surrounding healthcare access, employment rights, and personal identity will become increasingly contentious as these "post-human" individuals navigate a legal system that may not recognize their unique circumstances.

Moreover, the potential for discrimination against those who defy traditional human norms raises ethical dilemmas regarding inclusion and societal values. If the law does not evolve to accommodate these new forms of human existence, we risk creating a class of citizens in legal limbo—biologically alive but politically disenfranchised. Such a bifurcation of society could create a situation where "full" citizens enjoy rights and protections, while "post-human" individuals are relegated to the status of *zoe*, stripped of the protections afforded to legitimate participants in the political community. This dilemma also applies to vulnerable populations who may not have chosen their status. Genetically modified organisms or individuals conceived through advanced reproductive technologies may find themselves subjected to societal judgment that questions their humanity. These issues complicate the conversation surrounding rights and recognition as society grapples with the ethical ramifications of biotechnological innovation and its effect on our understanding of citizenship.

Transhumanism, a philosophical and technological movement aiming to transcend human limitations—including mortality—introduces yet another layer to this conversation. Transhumanists envision a future where biological constraints no longer limit human potential. The dream of eternal life, once relegated to the realm of myth, is becoming a real possibility for the wealthy elite. Silicon Valley has become a hotbed for funding life-extension technologies, cognitive enhancements, and even digital immortality. Genetic manipulation, artificial organs, and brain-computer interfaces are all part of the drive to extend human life indefinitely (Kurzweil, 2005). As biotechnology, artificial intelligence, and robotics converge, the prospect of transcending mortality becomes a tangible, albeit controversial, aspiration.

However, transhumanism brings with it deep ethical questions. If we can extend life indefinitely, what does that mean for our understanding of the human condition? The traditional experiences of aging, death, and the finite nature of life shape how we find meaning and purpose. If mortality is no longer inevitable, does this change the way we view life itself? Moreover, as only the wealthy may afford these life-extending technologies, society faces the prospect of a growing divide between those who can afford immortality and those who cannot, raising questions of fairness, inequality, and justice (Fukuyama, 2002). The potential for a future in which only a privileged few live indefinitely forces us to confront the ethics of access and fairness.

As biotechnology continues to advance, traditional frameworks of human identity and citizenship are being redefined. Cloning, hybridization, and the transhumanist dream of transcending the human condition challenge not only our biological understanding but also our social and legal systems (Haraway, 1991). As artificial enhancements reshape what it means to be human, new questions arise about whether existing legal systems can accommodate the diversity of identities and experiences emerging in the 21st century. The rise of biotechnology compels us to reconsider fundamental aspects of human existence and rethink the societal structures that regulate it.

CHASING COPIES

Cloning, the scientific marvel of crafting genetically identical organisms, is a process where the clone carries the exact same DNA sequence as its original counterpart. While the idea of cloning may seem like something straight out of science fiction, its origins stretch back to the early 20th century. Early experiments, especially with amphibians, laid the foundation for today's advanced cloning technologies. The breakthrough moment came in the 1950s when scientists achieved the first success in cloning frogs. By transferring the nuclei from adult frog cells into enucleated eggs, they provided the first definitive proof that cloning from adult cells was not only possible but could become a reality (Briggs & King, 1952).

However, it wasn't until 1996 that cloning truly captured global attention with the birth of Dolly, the first mammal cloned from an adult somatic cell. This momentous event proved that adult mammalian cells could be reprogrammed to generate a new organism, opening a door to possibilities both thrilling and unsettling (Wilmut et al., 1997). Dolly, born from a mammary gland cell of a six-year-old sheep, marked a stark contrast to previous cloning efforts that relied on embryonic or fetal cells. Using somatic cell nuclear transfer (SCNT), scientists transferred the adult cell's nucleus into a denucleated egg cell, which was then stimulated to divide and develop into an embryo, ultimately implanted into a surrogate mother. The result? A genetically identical sheep, a process that would soon make its mark on history (Wilmut et al., 1997).

Dolly's birth ignited not only scientific excitement but also fierce ethical debate. While the potential for medical advancements was enticing—cloning could pave the way for regenerative medicine, organ transplants, and treatment for genetic disorders—the ethical concerns were equally daunting. Religious groups, ethicists, and policymakers were left grappling with a question that had become impossible to ignore: what are the moral boundaries of cloning, particularly when it comes to humans? The idea of reproductive cloning, or the creation of a human clone, raised alarms about the commodification of human life and the dehumanization that could follow, including the creation of a subclass of humans with fewer rights (Peters, 1997). In short, cloning stirred a conundrum that brought the tension between science and humanity into stark focus.

Since Dolly, cloning technology has continued to evolve. Scientists have cloned various animals, from cows and goats to mice and primates (Tachibana et al., 2012). This leap forward solidified cloning's place in the biological and biotechnological fields. However, while animal cloning has shown significant progress, human cloning remains speculative, a line scientists have not yet crossed due to myriad technical, ethical, and legal barriers.

Therapeutic cloning, which uses cloned embryos to harvest stem cells for medical treatment, has seen notable developments. Stem cells generated from cloned embryos hold promise for personalized medicine, potentially offering treatments customized to the individual's genetic makeup (Takahashi & Yamanaka, 2007). Yet, therapeutic cloning has also attracted its share of opposition, particularly due to the destruction of human embryos, which raises difficult questions about when life truly begins (Harris, 2004a, 2004b). Meanwhile, reproductive cloning, which would create an entire organism, remains illegal in many countries and is weighed down by technical challenges, such as high failure rates and concerns over the health of clones. Dolly herself, who lived only six years—half the typical lifespan for a sheep—suffered from premature aging and arthritis, sparking fears that clones could experience similar frailties (Shin et al., 2002).

In the present day, cloning research remains vibrant, particularly within agriculture and medicine. In agriculture, cloning allows for the reproduction of animals with desired traits, such as disease resistance or increased milk production (Huang et al., 2010). In medical research, cloning continues to be explored for the creation of genetically identical animals for controlled experiments. Moreover, the potential of cloning extends beyond current species to the realm of conservation. In 2020, scientists successfully cloned a black-footed ferret, a critically endangered species, marking an exciting leap forward in the use of cloning for biodiversity preservation (Feldman et al., 2021).

While human cloning remains a distant prospect mired in ethical and technical dilemmas, the rapid pace of technological advancement pushes the limits of possibility. The conversations surrounding cloning—whether applied to animals, humans, or extinct species—force us to rethink our understanding of biology, ethics, and what it means to replicate life in an era defined by biotechnology. As cloning continues to evolve, it calls into question not only the potential for new life but also the moral and societal implications of reproducing life itself.

CLONING STATUS QUO

Cloning has evolved into two distinct scientific territories: therapeutic and reproductive cloning. Both paths have their own unique goals, methodologies, and ethical landmines, shaping the future of medicine and challenging our fundamental understanding of life.

Therapeutic cloning is the medical marvel in the making, designed to grow tissues or organs for transplantation and treatment. The process is as clever as it sounds—by using a patient's own DNA to create genetically identical cells, scientists hope to replace damaged tissues with a perfect match. This could unlock a world where organs are custom-made for each patient, reducing transplant rejection risks and sparking new hope for conditions like Parkinson's disease, diabetes, and spinal cord injuries (Lanza et al., 1999). If successful, therapeutic cloning could revolutionize the dire shortage of organ donors and take medical treatment to a new level of precision (Daley et al., 2010).

Reproductive cloning, however, is a different beast altogether. Its goal is to create an entire organism, using techniques like SCNT. While it's theoretically possible to clone a human, reproductive cloning has become a moral battleground. Critics argue it could reduce human life to a mere commodity, raising the specter of exploitation where clones are seen not as individuals, but as products (Peters, 1997). Even more troubling, cloning often has a poor track record—only a fraction of cloned embryos survive to full term, and even fewer make it without severe health complications, such as genetic defects and premature aging. This raises deep ethical concerns about the welfare of clones and the potential risks they might face.

Still, the science marches on. One of the brightest beacons in the cloning world is the development of induced pluripotent stem cells (iPSCs), which allow researchers to reprogram adult cells to an embryonic-like state. This innovation paves the way for regenerative medicine, offering a rich source of cells that can morph into various tissue types, all without the ethical baggage tied to embryonic stem cells (Takahashi & Yamanaka, 2007). The potential of iPSCs is staggering, offering new possibilities for drug development, disease modeling, and therapies for conditions that were once considered untreatable.

But even as cloning techniques evolve, the prospect of replicating a human being remains fraught with philosophical and ethical chaos. Many ethicists argue that cloning humans could erode the very essence of individuality and autonomy. If a clone is genetically identical to another person, does that mean they share the same personality, memories, or consciousness? And what of the societal implications? If clones are created, could they be relegated to a new, subhuman class—deprived of rights or exploited for labor? These unsettling questions further complicate the cloning debate.

As the science behind cloning advances, it's clear that we are standing at a crossroads between groundbreaking medical possibilities and profound ethical quandaries.

Therapeutic cloning could bring about unparalleled medical breakthroughs, but reproductive cloning forces society to reckon with difficult moral choices. As cloning technologies continue to evolve, we will need to strike a delicate balance between scientific progress and the values we hold dear about human life and dignity.

ETHICAL CROSSROADS

One of the most provocative questions raised by cloning is the very nature of identity and individuality. If a human can be perfectly replicated, what happens to the concept of uniqueness? A clone may share the same genetic makeup as its original, yet this raises profound questions about what it truly means to be "the same" person. The idea of a "perfect replica" is far from simple. While the clone might resemble the original physically and potentially inherit similar traits or health predispositions, their identities would inevitably diverge due to the influences of experience, environment, and social interaction.

This issue becomes even more complicated when we consider the nature vs. nurture debate, which suggests that both genetics (nature) and environment (nurture) shape an individual's personality and behavior. Clones, although genetically identical, wouldn't share the same lived experiences as their originals. For example, a clone raised in a different environment could develop distinct preferences, beliefs, and personality traits, leading to a different sense of self (Dawkins, 2006). This raises a critical question: if genetics alone can't account for individuality, what other factors contribute to the essence of a person? And if cloning becomes a reality, how will society navigate the definition of uniqueness and personal identity in a world where duplicates exist?

Cloning also stirs metaphysical dilemmas about the soul and consciousness. In many religious and philosophical traditions, the soul is seen as an irreplaceable element of human life—the essence of individuality and personhood. If a clone shares the same genetic material, could it possess the same consciousness or soul as the original? Some argue that consciousness is shaped by a complex interaction between biology and experiential learning, meaning a clone might develop a separate consciousness based on its unique experiences (Kant, 1781).

Moreover, this debate often leads to broader questions about the nature of existence itself. If a clone is artificially created, rather than naturally conceived, does it possess a soul? This echoes religious teachings that emphasize the sanctity of natural birth. Philosophers like Descartes have questioned whether consciousness is merely a product of physical existence or a distinct, immaterial essence that cannot be replicated through cloning (Descartes, 1641). This philosophical dilemma forces us to reconsider the very foundations of human existence and identity, challenging our understanding of what it means to be alive.

A further challenge is the legal status and rights of clones. If clones are created, do they possess the same right to life as naturally born humans? The current frameworks governing personhood are already complex, and introducing clones into society could further muddy these waters. Clones might face discrimination, social stigma, and legal ambiguity, raising ethical questions about how they should be treated (Harris, 2004a, 2004b).

Should clones have the same legal rights and protections as their originals, or would they occupy a different, potentially subordinate, class of beings? This is not just a theoretical issue—it has profound implications for how society defines personhood and human rights. If a clone is seen as a mere product of science, society may be inclined to treat it as less than human, risking exploitation and unethical treatment. These questions extend beyond individual rights, influencing how laws are crafted and how ethical standards evolve in response to rapidly advancing biotechnology (Bostrom, 2014). Navigating these dilemmas requires us to confront what it truly means to be human in a world where cloning is a potential reality. Each of these issues compels us to re-examine the foundations of identity, consciousness, and human rights, forcing a broader reconsideration of our ethical and legal systems.

One of the most contentious issues surrounding cloning is whether a clone, as a perfect genetic replica of a human, should be granted the same legal status and rights as any other individual. This raises vital questions about the nature of human rights: does the creation of a clone automatically endow it with the rights to life, liberty, and personhood? Some argue that if clones possess human DNA and exhibit characteristics associated with humanity—such as emotions, reasoning, and social interaction—they should be granted the same rights as naturally conceived humans (Waldby & Mitchell, 2006).

On the other hand, critics argue that clones are inherently different because they do not emerge from the natural biological process of conception. This complicates the question of personhood: is it determined solely by genetics, or is there an intrinsic quality—such as consciousness or soul—that defines one's legal and ethical status? Currently, legal frameworks around the world do not recognize clones as full persons, leaving them in a state of ambiguity regarding their rights and protections under the law (López, 2014).

THE LAW ON CLONING

The regulation of cloning is a patchwork of policies, varying wildly across the globe, shaped by a mix of cultural, ethical, and religious values. In many countries, including the United States, human cloning is met with strict opposition or outright bans, largely driven by ethical concerns. In the United States, the Dickey-Wicker Amendment ensures federal funds cannot be used for research that involves creating or destroying human embryos, effectively stalling reproductive cloning efforts and placing substantial roadblocks in the path of research and development (Harris, 2004a, 2004b).

Meanwhile, in the United Kingdom, therapeutic cloning is allowed—but only under tightly controlled regulations designed for medical research. Scientists can create cloned cells to study diseases and develop treatments, but reproductive cloning is strictly forbidden. This reflects a cautious, wait-and-see approach to the potentially perilous consequences of cloning entire organisms (Keenan, 2009).

Of course, the ethical debate surrounding cloning is not just academic—it's deeply intertwined with religion. Many religious groups vehemently oppose cloning, arguing it violates the natural order or undermines the sanctity of human life. The Catholic Church, in particular, has been vocal in its disapproval, claiming that cloning diminishes the dignity of human procreation (Pope John Paul II, 1998). These religious objections have been instrumental in shaping international laws, with some countries opting for outright bans while others allow controlled research to proceed, albeit under strict conditions.

The United Nations has also stepped into the fray, grappling with the ethical implications of cloning. In 2005, a U.N. declaration called for a global ban on all forms of human cloning that contradict human dignity and the protection of life. However, since the declaration is non-binding, its impact has been limited, and enforcement varies widely across nations (United Nations, 2005). This lack of a unified global stance underscores the complexity of the issue and highlights the difficulties of reaching a consensus on the ethical and legal status of cloning.

As cloning technology continues to evolve, the political, legal, and ethical hurdles will likely shift and intensify. The prospect of human cloning forces us to confront some of life's most fundamental questions: What does it mean to be human? What is the value of identity and individuality in an age where replication is possible? The future of cloning will be shaped by ongoing debates that strive to balance the potential benefits of this technology with the deep philosophical and moral questions it raises, challenging us to rethink the very essence of humanity itself.

MODERN HYBRIDS

The modern landscape of body modification has undergone a seismic shift, with extreme surgeries, cutting-edge prosthetics, and bio-hacking technologies taking center stage. This evolution isn't just about tweaking appearances—it's a full-scale challenge to the very notion of what it means to be human. As technology becomes more accessible, people are increasingly pushing the boundaries of their bodies, not just for aesthetic reasons, but to enhance physical abilities, redefine gender identity, and even incorporate cybernetic elements into their very biology.

Extreme body modifications now span a vast spectrum of practices, from implants designed to amplify human capabilities to complex surgeries that radically alter one's appearance. These modifications often include cyborg-like enhancements that improve strength, agility, or sensory perception, allowing individuals to push past the limits of biology. Advanced prosthetics, for instance, have enabled amputees to regain mobility and function, now with added features like electronic sensors and robotics, transforming disability into a new realm of human possibility (Sparrow, 2014). Beyond physical enhancements, gender modifications are rewriting the script on gender identity. Gender reassignment surgeries and hormone therapies empower individuals to align their physical selves with their gender identities, embracing a view of gender as a fluid spectrum, rather than a rigid binary. This isn't about fitting into societal molds; it's about reclaiming authenticity and self-expression in a world that increasingly questions traditional identities.

Body modification itself is far from new. Take tattoos—throughout history, they've served as symbols of tribal belonging, spiritual belief, and social status. In many cultures, tattoos have been powerful markers of identity and resilience. Ancient Polynesian tattoos, for example, were deeply embedded in cultural identity, signifying social rank, achievements, and personal stories. Similarly, Native American tribes used tattoos to mark pivotal life moments, with each design telling a piece of the wearer's personal journey.

But the perception of tattoos has been anything but static. In the 19th and early 20th centuries, tattoos were often viewed through a lens of marginalization, closely associated with sailors, criminals, and circus performers—people on the fringes of society. Tattooed individuals were often seen as "others," struggling to fit into a society that deemed them deviant (Mifflin, 1997). Fast forward to today, and tattoos have become mainstream, embraced as a legitimate form of self-expression. This shift in attitude parallels the broader acceptance of extreme body modifications, hinting at a cultural trajectory toward greater inclusion and personal autonomy.

What's driving this movement forward is not just cultural change, but also breakthroughs in medical technology. Advances in surgical techniques, materials science, and bioengineering have made extreme body modifications safer and more accessible. These innovations have opened up new avenues for exploration, allowing people to reimagine their bodies and identities in ways once confined to science fiction. At the same time, society's increasing acceptance of diverse expressions of identity has given individuals the freedom to see these modifications not as acts of conformity, but as bold statements of personal agency (Bennett, 2015).

This shift reflects a broader cultural reevaluation of identity, one where the lines between what it means to be human are no longer clearly drawn. Today, modern hybrids—a fusion of human biology and advanced technology—embody a more fluid understanding of self. They transcend traditional definitions of gender, race, and ability, embracing a world where personal choice and technological enhancement define the individual. In this evolving landscape, body modification becomes an act of defiance against societal norms, a celebration of human potential, and a bold declaration of who we can become (Haraway, 1991).

SOCIOCULTURAL BACKDROP

The cultural drive for perfection, bodily autonomy, and individual expression is undeniably fueling the rise of body modifications. This trend, however, isn't born from a sudden impulse—it has deep roots in earlier forms of self-expression, like tattoos and piercings, which have long been used as personal signatures and markers of identity. These practices have historically allowed people to carve out a space for themselves in a world often bent on uniformity, giving them a sense of control over their bodies (Klein, 2013). Tattoos, in particular, have served as symbols of status, rites of passage, and personal stories, weaving deeper cultural meanings into the very fabric of one's skin (Thompson, 2019).

As society loosens its grip on rigid expectations and opens up to a broader spectrum of identities, extreme body modifications are gaining ground as legitimate forms of self-expression. This shift represents a larger cultural transformation, one where bodily autonomy is not just a concept but a celebrated reality—where individuals have the freedom to sculpt their physical forms according to their deepest desires. The growing recognition of diverse gender identities and expressions, for example, has paved the way for increased acceptance of gender modification surgeries, allowing individuals to align their bodies with their inner selves (Budge et al.,

2013). Here, body modifications are more than cosmetic alterations; they are deeply connected to the pursuit of authenticity and self-realization.

What's also driving this change is the expanding presence of modified bodies in pop culture. Whether through influencers on social media or characters on TV, these modified bodies are being normalized, making them increasingly mainstream. Platforms like Instagram and TikTok have become virtual catwalks where people can flaunt their modified bodies, share their personal stories, and forge communities around shared experiences (Duffy & Hund, 2015). These digital spaces have become havens for those who feel sidelined by conventional beauty standards, giving rise to a more inclusive understanding of beauty. Consequently, extreme body modifications are transforming from signs of rebellion to declarations of individuality.

But it's not just cultural acceptance that's at play here. The rise of the body modification industry—spanning cosmetic surgery clinics, body art studios, and wellness centers—has provided a wide range of options for those looking to express themselves through their bodies. With so many avenues available, one must wonder: Are these modifications truly driven by personal desire, or is there an underlying pressure to conform to societal ideals of beauty and perfection? This tension brings to the surface broader issues of body image, self-worth, and the weight of media-driven expectations (Tiggemann, 2015).

BREAKING BIO

The rise of body modifications beckons a rethinking of traditional views on human limitations and identity. As more individuals alter their bodies with surgeries, prosthetics, and bio-hacks, they challenge the age-old notion of a fixed human essence. This shift stirs deep philosophical inquiries: Are we solely defined by our biological makeup, or is our identity a fluid construct, shaped by our choices and experiences? The conversations surrounding body modifications advocate for a break from the static conception of humanity, championing a dynamic understanding of identity that accommodates both biological and technological influences (Haraway, 1991).

This philosophical journey intersects with posthumanism, which suggests that the boundaries of what it means to be human are increasingly porous. Transhumanist thought invites us to rethink human identity in the context of technology's rapid advance, where digital and biological elements blur into one. Take cyborg implants or neuro-enhancements, for example—these modifications prompt questions about cognition, agency, and social interaction (Braidotti, 2013). As individuals embrace alterations to their physical and cognitive capacities, the line between human and machine grows increasingly indistinct, redefining what it means to be human in the process.

At the heart of the body modification movement is the emphasis on personal agency as a key force in shaping identity. By modifying their bodies, individuals take charge of their autonomy, challenging societal norms about appearance and functionality. This assertion of agency fosters a more inclusive view of identity—one that embraces the diversity of human experience and the many ways in which individuals can express themselves (Foucault, 1988). From this perspective, identity isn't a fixed essence; it's a fluid, ongoing process of self-creation and constant reinvention.

The body-as-canvas concept underscores the potential for perpetual transformation, both aesthetically and functionally. This perspective aligns with postmodern theories of identity, where the self is seen as ever-changing and adaptable. Just as an artist uses a canvas to manifest creativity, individuals can use their bodies to convey their personal narratives. This philosophy suggests that the body is not a limitation but rather a dynamic medium for self-exploration and expression, imbuing the act of modification with empowerment. Body modifications—whether tattoos, piercings, or surgical alterations—serve as visual statements that communicate milestones, beliefs, and affiliations, reflecting the personal and cultural stories they carry. In this light, modification becomes an art form, with the body constantly redefined, much like a living masterpiece, and the process of transformation integral to one's journey of self-discovery and self-actualization.

Furthermore, the body-as-canvas perspective challenges the traditional beauty standards imposed by society. By embracing modifications, individuals resist the pressure to conform to narrow ideals of aesthetics, asserting their identities on their own terms. This resistance fosters a sense of solidarity among those who modify their bodies, creating communities where alternative forms of beauty are not just accepted but celebrated. Ultimately, seeing the body as a canvas empowers individuals to reclaim their bodies as spaces for personal expression and creativity, taking ownership of their physical forms and the narratives they tell.

IDENTITY BATTLES

The question of human identity becomes deeply political when individuals modify their bodies to the point of being perceived as "non-human." This transformation carries significant implications for political, legal, and social rights. If a modified individual deviates so drastically from societal norms that they are no longer recognized as fully human, they risk losing their legal personhood. These concerns force us to confront the fragility of identity within political discourse and the potential consequences of extreme bodily alterations.

The implications of body modifications stretch far beyond personal choice, intersecting with broader societal perceptions of normativity and deviance. When individuals undergo modifications that challenge conventional definitions of humanity, they may face legal ambiguities regarding their status. In many jurisdictions, legal personhood is still tied to specific biological and social criteria that reflect traditional notions of humanity. If someone's modifications push them into the "non-human" category, they may lose rights typically afforded to humans, such as voting, access to healthcare, and legal representation (Bennett, 2015).

This looming loss of rights compels us to reconsider what it means to be recognized as human in today's society. It conjures up fears of discrimination and marginalization, where those who embrace radical body modifications may be relegated to the status of "others," stripped of full citizenship rights. This scenario echoes historical precedents where marginalized groups—such as those with disabilities or non-conforming gender identities—have fought for recognition and rights in the face of societal norms (Davis, 2015). The fear of being deemed non-human creates a chilling effect on personal expression, with individuals hesitating to modify their bodies for fear of social and legal repercussions.

RIGHTS OF NON-HUMANS

As body modifications push the boundaries of what it means to be human, exist-ing laws strain to keep up, creating a legal landscape fraught with ambiguity. The potential legal status of highly modified individuals raises contentious questions about citizenship and human rights protections. For instance, if society begins to classify individuals with extreme modifications—such as extensive prosthetics, bio-engineered alterations, or significant enhancements—as non-human, they risk losing fundamental rights. These rights could include not only citizenship but also legal protections traditionally reserved for human beings (Bennett, 2015).

This possible shift prompts critical questions about how personhood is defined within legal frameworks. Legal systems, traditionally rooted in biological and social norms, may struggle to accommodate individuals whose bodies deviate from the established standard of humanity. In some jurisdictions, laws are explicitly tied to traditional views on what constitutes a human being, including biological criteria and social responsibilities. As such, those who undergo radical body modifications may find themselves in a precarious position, excluded from the rights and privi-leges enjoyed by typical citizens (Davis, 2015). This issue is especially pressing when modifications are perceived as reducing an individual's humanity, resulting in dis-crimination and social ostracism.

The ambiguity surrounding the legal definitions of personhood presents signifi-cant challenges for lawmakers and society at large. As body modifications become more widespread, legal systems must grapple with the implications and consider whether existing laws are sufficient to protect the rights of all individuals, regardless of their physical appearance. If the legal status of "non-humans" remains undefined, it could set a dangerous precedent, making highly modified individuals invisible in the eyes of the law and further marginalizing them in society (Agamben, 1998).

The issue of self-identification presents another critical dilemma in the debate surrounding body modifications. Can a person legally identify as "non-human"? If laws fail to recognize modified individuals as fully human, they may face substantial challenges in asserting their identities. This legal gap raises concerns about the right to self-identification and its implications for personal agency.

Self-identification is a cornerstone of personal autonomy, allowing individuals to define their own identity and existence. But when societal and legal frameworks lag behind evolving understandings of identity, modified individuals may find their self-identification invalidated or questioned. For example, legal systems may refuse to allow someone to change their legal identity to reflect their modified status, cre-ating conflict between personal beliefs and legal recognition (Harris, 2004a). The struggle for recognition as non-human could provoke resistance from legal systems that are slow to adapt to new definitions of identity, leading to further marginaliza-tion and alienation for those who don't fit neatly into traditional categories.

Moreover, the legal recognition of self-identified non-human individuals could spark broader societal discussions about the nature of humanity and the rights that come with it. If a modified individual successfully asserts their right to be recognized as non-human, it could challenge prevailing ideas of citizenship and rights in ways that demand legal reform (Bennett, 2015). However, such a challenge would likely meet significant resistance, including moral objections rooted in religious beliefs or societal norms that reject the idea of non-human identities.

BIAS AND BACKLASH

The rise of extreme body modifications stirs up significant social and political resistance, tapping into deeper societal anxieties about identity, ethics, and the limits of human expression. As individuals embrace radical alterations to their bodies—whether through tattoos, piercings, surgeries, or even technological enhancements—they often find themselves at the center of intense debates. These discussions typically pivot around medical ethics, religious beliefs, and moral opposition, which can breed discrimination against those who choose to modify their bodies. Critics argue that extreme modifications challenge societal norms, potentially setting troubling precedents regarding human experimentation and ethics (Bennett, 2015).

This backlash can manifest in a range of forms, from legislative efforts to restrict body modifications to the social stigmatization of those who undergo such changes. For example, certain jurisdictions have proposed laws regulating or outright banning specific modifications, citing safety concerns and the potential for exploitation. These legislative moves often come from a paternalistic standpoint that assumes individuals lack the capacity to make informed choices about their own bodies, thereby undermining personal autonomy.

In addition, societal stigmatization can surface as discrimination in areas such as employment, healthcare, and everyday social interactions. People with visible body modifications may find themselves subject to biases that leave them vulnerable to exclusion or marginalization. Research shows that individuals with extreme body alterations often face negative judgments, with their professionalism or reliability called into question. Such discrimination not only impacts individuals on a personal level but also reinforces broader societal norms that favor conventional appearances and behaviors, sidelining diversity.

TRANSHUMANISM

Transhumanism emerges as a provocative philosophical movement that envisions a redefinition of the human condition through the power of advanced technologies. Grounded in the belief that humanity can transcend its physical and mental limitations, transhumanism suggests that, through scientific innovation and technological progress, we can radically alter our experience of life, pushing past the traditional boundaries of human existence. The ultimate goal of this transformation is immortality—an idea that raises profound questions about the very essence of humanity, the nature of life, and the ethics surrounding such advancements (Bostrom, 2014).

Transhumanism's roots are deep, entwined in both ancient philosophies and modern technological advancements. This section delves into key figures in transhumanist thought, exploring the contrasts between religious beliefs about immortality and the contemporary, tech-driven pursuit of life extension.

Max More and Ray Kurzweil are two towering figures in the development of transhumanist thought. Max More, philosopher and futurist, helped articulate the foundational principles of transhumanism in the 1990s. His essay, "The Proactionary Principle," advocates for a proactive stance toward technological innovation, highlighting the potential benefits of biotechnological advancements while acknowledging the risks involved (More & Vita-More, 2013). More's emphasis on enhancing life through technology aligns with the core tenet of transhumanism: that humanity can rise above its biological limitations.

Ray Kurzweil, a renowned inventor and futurist, further popularized transhumanist ideas through his books and predictions about the future of technology. In *The Singularity Is Near* (2005), Kurzweil argues that humans will eventually merge with machines, leading to a point of "singularity" where technological growth becomes uncontrollable and irreversible. This convergence, he suggests, will not only enhance human cognition and health but could even make immortality a reality through digital consciousness. Kurzweil's optimistic vision of the future has significantly shaped public discourse around transhumanism, promoting the idea that technology holds the key to solving humanity's fundamental challenges—death included.

Immortality, a perennial theme in many religious traditions, is also central to transhumanism. Most religions offer narratives of eternal life or an afterlife, framing death as a transition to another state of being. Christianity speaks of the resurrection and eternal life in Heaven, while Hinduism presents reincarnation as an endless cycle of birth, death, and rebirth, ultimately leading to Moksha, or liberation. These religious views provide ethical guidelines for understanding life, death, and existence, often emphasizing virtuous living in accordance with divine laws.

In contrast, transhumanism takes a technological approach to immortality, seeking to conquer death through scientific advancements. Proponents argue that by employing technologies like genetic engineering, nanotechnology, and artificial intelligence, humanity can overcome biological constraints and achieve immortality—not in a metaphysical sense, but through physical means (Fukuyama, 2002). This materialist pursuit of immortality often rests on the assumption that human consciousness is a biological product, capable of being replicated or enhanced by technology.

The divergence between religious and technological immortality raises serious ethical dilemmas. While religious doctrines typically offer moral frameworks for attaining eternal life, technological immortality raises concerns about access: who will be able to afford these life-extending technologies? Moreover, reducing life and death to mere technological processes risks stripping away the spiritual and existential dimensions that have traditionally defined these concepts, challenging long-held beliefs about life's meaning and purpose.

In essence, transhumanism is motivated by a desire to overcome mortality. This pursuit encompasses various technologies and philosophies aimed at extending life indefinitely. The basic idea is that death—long viewed as an inevitable part of the human experience—should be confronted, not accepted. Advances in genetics, biotechnology, and aging science have driven transhumanists to advocate for applying this knowledge to enhance human life, allowing for not just a longer lifespan but a better quality of life. For instance, gene-editing technologies like Clustered Regularly Interspaced Short Palindromic Repeats (CRISPR) could potentially eliminate age-related diseases, ushering in a new era of longevity (Doudna & Charpentier, 2014).

The quest for immortality is rapidly becoming a central topic across academic and professional fields. Ethically, the moral implications of extending life present serious challenges. Critics argue that manipulating life and death in such a way undermines the natural order, leading to unforeseen dilemmas (Fukuyama, 2002). For example, if humanity succeeds in extending life dramatically, what responsibility do we have toward future generations? Should the pursuit of immortality be restrained by ethical limits, and who decides where those boundaries lie?

Philosophically, the prospect of escaping death raises profound questions about the meaning of life. If death is no longer a certainty, what happens to our understanding of purpose and fulfillment? Traditional views often derive life's significance from the inevitability of death, but the potential for eternal life complicates this narrative. Some philosophers argue that without death, life could lose its urgency and depth (Sandel, 2007). Existentialist perspectives in particular highlight the tension between the desire for immortality and the search for meaning, asserting that our fear of death fuels our quest for significance.

TRANSHUMANISM TECH

Transhumanism is driven by a suite of emerging technologies that promise to radically transform human life and experience. At its heart are advancements in biohacking, genetic engineering, cryonics, and mind uploading, along with hefty investments from tech industry titans in life-extension ventures. These technologies not only reflect the bold aspirations of transhumanists but also raise profound ethical and philosophical questions about humanity's future.

Biohacking and genetic engineering lie at the forefront of transhumanism's quest to conquer aging and disease. One of the most transformative tools in this realm is CRISPR , a game-changing gene-editing technology that allows precise modifications to DNA. CRISPR enables scientists to edit genes with astounding accuracy, potentially eliminating genetic disorders, enhancing physical traits, and even slowing the aging process (Doudna & Charpentier, 2014). For example, researchers have already experimented with CRISPR to target genes linked to diseases like sickle cell anemia, showing its potential for long-term cures.

However, the potential of CRISPR raises crucial ethical concerns. Critics worry about the advent of "designer babies," where parents could select traits for their offspring, creating societal disparities and dilemmas around consent and fairness (Sandel, 2007). While the technology offers hope for improved health and longevity, it also forces us to confront the moral responsibilities inherent in wielding such power.

Cryonics offers another intriguing aspect of transhumanism: the preservation of individuals at ultra-low temperatures after death, with the aim of reviving them when medical technology advances enough to cure their ailments. Cryonics advocates argue that, while current medical knowledge can't save certain patients, future technologies might offer them a second chance at life. Institutions like the Alcor Life Extension Foundation provide cryopreservation services, emphasizing the possibility of a future where aging and disease are vanquished.

Parallel to cryonics is the concept of mind uploading, which involves transferring a person's consciousness and memories into a digital form. Though still largely theoretical, this idea raises deep philosophical questions about identity and the continuity of self. If consciousness can be replicated or transferred, what does it mean to be human? Would a digital version of a person remain the same, or simply be an imposter?

THE FOUNTAIN OF TECH YOUTH

In recent years, Silicon Valley has transformed from the land of tech startups and innovation into a high-stakes laboratory for life-extension experiments. What was once the stuff of science fiction—immortality—now appears to be on the brink of becoming a marketable product, thanks to the likes of Jeff Bezos, Peter Thiel, and Larry Page. These billionaires are not just chasing the next big thing—they're chasing life itself. And in doing so, they're turning the very nature of aging and mortality on its head, creating a world where the idea of living forever is no longer a fantasy but a very real pursuit backed by scientific dollars.

Jeff Bezos, Amazon's founder and the man behind one of the most successful business empires of all time, has bet big on Unity Biotechnology, a company dedicated to attacking aging at the cellular level. Their strategy? Eliminate senescent cells—those stubborn, aging cells that refuse to divide and wreak havoc on our health. The science behind this is simple in theory but revolutionary in practice: removing these cells has shown promise in improving health outcomes in animal models. If it works, Bezos won't just be adding years to his life—he'll be leading a charge to revolutionize society's entire relationship with aging. Unity Biotechnology isn't just a business venture; it's Bezos's bet that he can live forever—or at least longer than the rest of us.

Meanwhile, Peter Thiel, the PayPal co-founder who's made a career out of disruptive thinking, has staked his claim on life-extension technologies with a portfolio that reads like a who's-who of immortality startups. One of his more controversial investments is Ambrosia, a company that draws on the ancient myth of youth and rejuvenation—by literally transfusing young plasma into older individuals. It sounds like a plot twist from a B-grade sci-fi movie, but it's one that Thiel is convinced could revolutionize aging. But he's not stopping there: Thiel also backs Calico Labs, a subsidiary of Alphabet (Google's parent company), which is taking on the biological puzzle of aging in its own groundbreaking way. Thiel's enthusiasm for transhumanism is no secret—he has publicly declared his goal to "defeat death," challenging the limits of what can be achieved in the name of longevity. For Thiel, it's all part of a bigger vision where technology isn't just improving lives—it's extending them indefinitely.

Then there's Larry Page, the co-founder of Google, who's thrown his weight behind Calico Labs. But this isn't just any tech project—it's a moonshot to outsmart death itself. Page's ambitions for Calico go beyond longevity: his goal is to fundamentally alter the way we understand aging and develop technologies that can push human lifespans to unimaginable lengths. In fact, Calico's collaboration with top scientists and academic institutions isn't just a casual partnership—it's a serious attempt to rewrite the rulebook of biology. The big names and big money behind Calico are ensuring that longevity research gets the kind of attention it has long lacked, making this venture one of the most high-profile examples of Silicon Valley's obsession with immortality.

The rush of investments from these tech moguls reflects a broader shift within Silicon Valley itself—a place that has always prided itself on solving problems but is now focusing on a challenge far more personal: the challenge of death. Aging is increasingly seen not as an inevitable part of life but as a treatable condition, much like any disease. The idea that we might one day halt, reverse, or even completely sidestep the aging process is gaining serious traction. And it's not just about adding years to your life—it's about adding quality, vitality, and health. Human Longevity Inc. (HLI), co-founded by genomic pioneer Craig Venter, is leading the charge here. By using cutting-edge genomic data, HLI is tailoring personalized treatments to extend healthy lifespans. If this technology pans out, it could mark the beginning of a new era where health is no longer dictated by the march of time but by the precision of science.

But before you start dreaming of a world where death is optional, it's crucial to ask: who gets to live forever? This brave new world of life extension comes with some glaring ethical dilemmas, and the biggest one might be inequality. While Silicon Valley's immortality projects are undoubtedly transformative, they also carry a risk: only the wealthy may be able to access these life-extending technologies. Just imagine a future where those with deep pockets are living longer, healthier lives, while the rest of us continue to age and die as we always have. The technology itself could become an exclusive club, with only the fortunate few getting an all-access pass to an extended life. This could create a society sharply divided between those who have access to immortality and those who do not, deepening social and economic inequalities.

The broader societal implications of these immortality projects are profound. If only the elite have access to life-extension technologies, we could find ourselves in a world with a new class of "enhanced" humans—individuals who have lived longer, healthier lives, and who possess a radically different set of experiences and opportunities than the general population. These disparities could redefine what it means to be human, creating a new underclass of those excluded from the scientific advances that promise to prolong life. Imagine a world where the wealthy live for centuries, while the rest of us are stuck with our normal, mortal timelines.

And then there's the question of meaning. If we conquer death, what happens to life? Traditionally, our mortality infuses everything we do with a sense of urgency and purpose. But if you live forever—or even just for centuries—how do you find meaning in your day-to-day existence? The fear of death has always been a powerful motivator in human life, but if that fear is alleviated by technology, will we lose the drive to live fully? These are not just ethical questions—they are existential ones. The rise of immortality, as glamorous as it sounds, could fundamentally alter our understanding of what it means to live a good life.

So, as Silicon Valley pours billions into the quest for eternal life, we must pause and consider: is immortality a gift—or a curse? In a world where death no longer holds sway, what new challenges will we face—and who will be the ones to face them?

THE WHAT-IFS

As transhumanism gathers steam, it raises profound philosophical questions about the meaning of life, the essence of humanity, and our quest for immortality. The merging of technology and existence forces a reevaluation of traditional beliefs and challenges our understanding of what it means to live a meaningful life.

A central question in the pursuit of immortality is how it alters our understanding of purpose and fulfillment. If death is no longer a certainty, the urgency to live a meaningful life may dissipate. Traditionally, life's finitude motivates individuals to seek significance, form relationships, and create legacies (Sandel, 2007). The knowledge that time is limited fuels the drive to accomplish and connect.

But if immortality becomes achievable, what happens to the quest for meaning? Viktor Frankl, the neurologist and psychiatrist, argued that meaning is derived from suffering, love, and the pursuit of goals. He believed life's significance often emerges from challenges (Frankl, 2006). If suffering and struggle can be mitigated or eradicated, would meaning itself still hold? A life without mortality could lead to ennui and a sense of purposelessness, undermining the very values that make life worth living.

The possibility of immortality also forces us to reconsider the essence of humanity. If humans can live forever, what does that mean for our understanding of existence? Martin Heidegger argued that awareness of mortality is integral to understanding life's temporality and our authentic selves (Heidegger, 1962). Removing mortality from the equation could fundamentally alter the human experience.

Michel Foucault explored how modern societies shape human subjectivity, arguing that social constructs and power dynamics shape our identities (Foucault, 1988). In a world of transhumanism, where individuals can modify their bodies and minds through technology, the shared human experience may dissolve, leading to fragmented identities and an increasingly alienated existence.

Friedrich Nietzsche's philosophy offers profound insights into the ethical and existential dimensions of transhumanism, particularly through his concepts of the *Übermensch* and Eternal Return. Nietzsche's ideas challenge contemporary views on human enhancement and immortality, prompting a deeper examination of what it means to live a fulfilling life. The *Übermensch*, introduced in *Thus Spoke Zarathustra*, represents an ideal individual who transcends conventional morality and societal norms to create their own values and meaning (Nietzsche, 1883). Transhumanists' pursuit of immortality and enhancement reflects a contemporary quest for the *Übermensch*—seeking to surpass human limitations and redefine existence.

Yet this pursuit must be tempered with caution. The desire for radical enhancement may lead to a disconnection from the struggles that contribute to personal growth. Nietzsche's philosophy emphasizes the importance of embracing life's challenges as integral to the human experience. The transhumanist yearning for immortality may risk overlooking the value of suffering and the depth it brings to existence.

Nietzsche's concept of Eternal Return further complicates the transhumanist vision. The idea challenges individuals to consider whether they would be willing to live their lives over and over, experiencing every joy, pain, and mundane moment for eternity (Nietzsche, 1883). The prospect of immortality poses the question: would living forever lead to stagnation, or would it provide endless opportunities for growth?

Transhumanists must grapple with the tension between striving for the *Übermensch* ideal and the existential weight of Nietzsche's Eternal Return. The pursuit of immortality may represent a rejection of life's cyclical nature, with its inherent struggles, in favor of escaping the limitations of mortality. Yet Nietzsche's call to live fully in the present, embracing both life's joys and hardships, reminds us that mortality's urgency is essential to human meaning.

PERILS OF BECOMING SUPERHUMAN

As society begins to merge advanced technologies with human biology, we're facing a Pandora's box of political and legal dilemmas surrounding transhumanism. These challenges prompt profound questions about inequality, legal status, and the very concept of rights in a post-human world.

One of the most glaring concerns is the potential for an ever-widening divide. As life-extending technologies become a reality, the wealthy will likely corner the market, turning immortality into a privilege for the elite. Picture a future where the "enhanced" class, with their superior health and cognitive abilities, hold an enormous advantage over their less fortunate, unenhanced peers (Bostrom, 2014). The future of humanity could become an exclusive club—one where entry is reserved for those with deep pockets.

But here's the rub: to keep these inequalities in check, we'll need regulatory frameworks that ensure equitable access to life-enhancing treatments. Policymakers must step up to the plate and design systems that provide universal access, much like the public health models that ensure basic services for all. However, a new question arises—how do we prevent a dystopian future where enhancements are monopolized by profit-driven interests? The ethical debates over patenting human enhancements and commercialization could set the stage for a new kind of class warfare, where only the wealthy can afford life-extending treatments (Sparrow, 2014).

And then there's the issue of "superhumans." As transhumanist modifications like brain enhancements become more common, we'll need to ask whether those who undergo these procedures will be recognized as a new class of beings entirely. Imagine a legal system where enhanced individuals, with superior abilities, face a unique set of rights and privileges that may not apply to the rest of humanity (Fukuyama, 2002). If enhancements truly push the boundaries of human capacity, the question will no longer be about whether they deserve rights but which rights apply to them.

This could spawn an entirely new legal framework—one that defines the rights of enhanced versus unenhanced individuals. But that could set off a ripple effect of discrimination claims and societal unrest over resource allocation. As humanity ventures into an era of biological enhancement, it's time to ask whether our existing human rights framework will suffice, or whether a "post-human rights" code is on the horizon.

Laws will soon need to grapple with defining rights for beings whose cognitive abilities, physical capabilities, and lifespans far exceed those of ordinary humans. The very notion of what it means to be human will need to be revisited. Can we still apply traditional human rights laws to these enhanced individuals, or is a rethinking of personhood in order? This opens the door to questions about the ethical treatment of individuals whose enhancements might render them more valuable—or more vulnerable—to exploitation by the state or corporate entities.

FINAL BYTE

As we race toward a post-human future, with groundbreaking advancements in technology and biological enhancements, society finds itself facing a buffet of social and ethical quandaries. These dilemmas are not just theoretical—they have real-world consequences that touch on everything from class divides to existential concerns about our humanity.

Take the rise of cloning, body modifications, and transhumanist enhancements. As these technologies develop, their costs will inevitably remain out of reach for most people, exacerbating already stark socio-economic divides. Imagine a world where life-extending treatments or advanced genetic modifications are available only to the rich, creating a society split between the "enhanced" elite, enjoying superior health and intellect, and the "unenhanced," struggling with age and health-related issues (Bostrom, 2014). This divide will affect more than just health outcomes—it will trickle into education, employment, and social mobility, as enhanced individuals gain a competitive edge. A two-tiered society could emerge, fueling resentment and social unrest among those left behind in the race for technological superiority.

Even more concerning is the potential for discrimination based on one's biological status. As transhumanist technologies become more pervasive, unmodified individuals may find themselves excluded or stigmatized in both social and professional spheres. A new form of caste system could emerge, with technological access determining one's place in society (Sparrow, 2016). This could lead to biased hiring practices, educational inequalities, and even social ostracism, further entrenching the divide between those with access to enhancements and those without.

Resistance to these technological advancements is already mounting, especially from religious and moral communities that see these developments as a direct challenge to established views on human life. For many, cloning and genetic engineering are viewed as sacrilegious—a violation of the sanctity of life. To these critics, the very idea of manipulating or creating life artificially runs counter to fundamental religious beliefs (Fukuyama, 2002). This resistance, though, will likely only grow louder as the technologies progress, spurring public campaigns, legal challenges, and calls for regulation or outright bans on certain enhancements.

Underlying much of the ethical opposition is the so-called "playing God" argument. Critics fear that we're taking a dangerous step by manipulating life at such a fundamental level. Could these advancements lead to unforeseen and possibly disastrous consequences? The moral questions raised by this new frontier force us to confront our responsibilities as stewards of life. Should we proceed, or is there a point where we must draw the line and accept our mortal limits?

Finally, there's the existential dilemma: what does it mean to be human in a world where we are enhancing our biology at an ever-increasing pace? Will we lose the very qualities that make us human—our empathy, creativity, emotional depth—in our quest for perfection? If we alter our bodies and minds, how do we even define ourselves anymore (Buchanan, 2011)? As we embrace enhancements, we risk sacrificing the very traits that bind us together as a species. This inquiry is central to the growing debate about how we balance technological advancements with preserving the core values that make us human.

As we integrate more enhancements into our daily lives, the fear is that we will lose our uniqueness. In our drive to create "better" versions of ourselves, we might overshadow the rich diversity that defines humanity. The rise of "ideal" beings could inadvertently erase the traits that make us who we are, raising critical questions about identity and individuality in a world where technology shapes the very core of our being.

These concerns are not confined to human enhancement alone; robotics, and the rise of humanoid machines, add another layer of complexity to our understanding of personhood and what it means to be human. In the next chapter, we will dive deeper into how advances in robotics are redefining our concepts of consciousness and individuality, and how the merging of humans with machines may change our understanding of what it means to be "alive."

6 Domo Arigato, Mr. Roboto

In 1983, the band Styx released "Mr. Roboto," a song that took the charts by storm, securing the number one spot for weeks. Its vibrant blend of synthesized sounds and thought-provoking lyrics swept listeners into a colorful world where technology and identity collided. The playful refrain "Domo arigato, Mr. Roboto" resonated like a friendly invitation, encouraging everyone to envision a future shaped by machines. For many, the song offered a glimpse into a world where technology was not merely a tool but a vital part of what it meant to be human. As the catchy melody played on radios and in homes, it sparked curiosity about the implications of this evolving relationship with technology. The lyrics served as a cultural mirror, reflecting a society grappling with rapid change. Amid the excitement of ATARI video games and robots on TV, a central question emerged: how do we define humanity when our identities might one day be echoed by the very machines we create?

This inquiry reverberated through various facets of popular culture, intertwining narratives that explore the complexities of humanity in an increasingly automated world. Iconic characters like R2-D2 and C-3PO from *Star Wars* (1977) exemplified the spectrum of robotic existence, blending loyalty, humor, and emotional depth while challenging our perceptions of intelligence and companionship (Bostrom, 2014). In Woody Allen's *Sleeper* (1973), the absurdities of a technologically driven future are satirized through the experiences of a man revived from cryogenic sleep. Navigating a landscape filled with bizarre machines and unconventional social norms, Allen critiques humanity's overreliance on technology while celebrating the resilience of the human spirit. This comedic portrayal reveals a truth that resonates: even in a world dominated by absurdity, the quest for connection and meaning remains deeply human.

Fast forward to today, and the exploration of robots and humanoids continues to evolve in a world where machines not only assist us but often mimic our very essence. This is exemplified in the 2014 film *Ex Machina*, which dives into the intricacies of artificial intelligence (AI). The story presents a young programmer tasked with assessing the emotional and cognitive capabilities of a humanoid robot named Ava. As Ava engages in conversations that blur the line between programmed response and genuine emotion, the film compels viewers to confront uncomfortable truths about autonomy, desire, and the nature of consciousness. The emotional stakes rise dramatically, pushing both the protagonist and the audience to grapple with the ethical implications of creating beings capable of mimicking and manipulating human feelings.

Contemporary series like *Humans*, *Westworld*, and *Sunny* amplify the discourse surrounding humanoids, portraying them as entities entwined in complex emotional and moral dilemmas. In *Humans* (2015), the titular synths are sophisticated

DOI: 10.1201/9781003624813-7

beings capable of experiencing emotions and engaging in moral reasoning. As these characters navigate a world filled with ethical complexities, the narrative invites audiences to explore the nature of companionship and the emotional bonds that can develop between humans and humanoids. Similarly, *Westworld* (2016) blurs the lines between creator and creation, raising questions about consciousness, free will, and the moral responsibilities of those who design sentient beings. These relationships fulfill psychological needs while complicating traditional notions of companionship, urging viewers to consider the ethical ramifications of forming attachments with artificial beings.

In *Sunny* (2024), the titular character is not just a highly advanced humanoid but also a being capable of emotional depth and moral reasoning. This narrative reflects the growing fascination with the potential for robots to transcend their mechanical origins and form meaningful connections with humans. As technology continues to evolve, humanoids like Sunny push the boundaries of what it means to be "alive" or to possess emotions. They serve as vessels through which we explore the evolving concept of companionship, inviting us to reconsider the roles robots may play in human life. The humanoid character embodies traits that were once thought to be uniquely human: empathy, love, and ethical decision-making. As the story unfolds, viewers are presented with a complex portrayal of a machine that does not simply follow orders but makes decisions based on emotional and moral considerations. This shift marks a significant departure from earlier portrayals of robots in media, which often depicted them as emotionless servants or destructive forces devoid of individuality.

Over time, robots in fiction have evolved from simplistic automata to sophisticated beings that evoke empathy. Early representations, such as the stiff, mechanical movements of robots in films like the 1927 German expressionist film *Metropolis*, offered cautionary tales about the potential dangers of automation. In contrast, contemporary representations emphasize the capacity for robots to form intimate bonds with humans, sometimes even eliciting stronger emotional connections than their human counterparts. One of the most fascinating aspects of these current representations is the emotional bond that develops between humans and humanoids. As humanoids become more sophisticated, they blur the lines between humans and machines, leading us to question the nature of emotional connections. Can these relationships be as fulfilling as those between humans? Do they cater to specific psychological needs that humans, for various reasons, may not meet with one another?

As robots gain more human-like qualities, cases have already emerged in which individuals form deep attachments to their humanoid companions, some even choosing to marry them (Levy, 2007). These cases, while rare, force society to grapple with the implications of such relationships. While it is easy to dismiss these cases as extreme or fringe, they highlight the extent to which humanoids can fulfill psychological needs—particularly those related to companionship, intimacy, and even love. As humanoids become increasingly able to "learn" from their human counterparts, these relationships could challenge traditional ideas of what constitutes meaningful companionship. This evolution has profound implications for how we view human relationships. If robots can provide comfort, loyalty, and emotional support, what does this say about our need for human companionship? As more people turn to artificial beings for emotional fulfillment, the lines between real and artificial connections become increasingly blurred, prompting questions about authenticity, intimacy, and the future of relationships in a highly automated world.

The central question guiding our exploration is: How do robots and humanoids redefine our understanding of what it means to be human in an increasingly automated world? This question invites us to consider the changing landscape of human-robot interactions and how the portrayal of humanoids in media reflects our shifting societal values. Humanoids like Sunny do more than serve as advanced technology—they provoke philosophical inquiry about what it means to possess consciousness, empathy, and morality.

Robots and humanoids in contemporary media challenge us to reflect on the qualities that define humanity. As humanoids become more advanced and lifelike, the question of what it means to be human is brought into sharper focus. Are we defined by our biology, or is it our ability to experience complex emotions and make moral decisions? When robots exhibit these same qualities, the distinctions between humans and machines grow murky (Bryson, 2010). Moreover, the societal implications of these portrayals extend beyond philosophical inquiry. Humanoids are increasingly being integrated into everyday life, from caretaking roles to companionship for the elderly. As this trend grows, we are forced to reckon with the ethical dimensions of human-robot relationships. What happens when the lines between creator and creation blur? How do we ensure that these relationships remain consensual and ethical, particularly when humanoids are programmed to fulfill human desires?

FROM TALOS TO ALEXA

The term "robot" was first coined by Czech writer Karel Čapek in his play *R.U.R. (Rossum's Universal Robots)*, where artificially created workers were designed to serve humans. The play, published in 1920 and premiered in Prague in 1921, is notable for introducing the term "robot," derived from the Czech word "robota," meaning forced labor or drudgery. In *R.U.R.*, these robots, initially created to serve humans, eventually rebel, leading to humanity's destruction. Čapek's work reflects early 20th-century anxieties about industrialization, addressing themes of exploitation, class struggle, and the dehumanizing effects of mass production. However, the conceptualization of humanoid robots dates back much further in human history.

The earliest representations of humanoid robots can be traced back to mythology. One notable example is *Talos*, the giant automaton from Greek mythology, crafted from bronze to protect the island of Crete. Talos was programmed to circle the island three times daily, defending it from invaders. While not a robot in the modern sense, Talos represents early ideas of creating non-human, artificial entities for specific tasks, highlighting humanity's enduring fascination with creation and control through technology. In Jewish folklore, the *Golem* serves as another example of an artificial being. Made from clay and animated through mystical means, the Golem was often portrayed as a protector or servant but also carried undertones of danger when it exceeded the creator's control. Both Talos and the Golem resonate with later robot narratives, where creations designed to serve humanity can ultimately become a source of destruction.

Philosophers such as Aristotle and Plato also touched on the concept of artificial beings, though in more abstract ways. In Plato's dialogue *Meno*, he suggested that knowledge could be implanted into individuals, akin to programming a machine (Plato, 1961). This metaphor implies a view of human cognition that aligns with later discussions of AI, positing that human learning and behavior could be shaped by external inputs, much like a machine's programming. Plato's emphasis on the transferability of knowledge raises fundamental questions about the nature of learning and intelligence, blurring the lines between human cognition and mechanical processes.

Aristotle, in his work *Politics*, delved into the potential of automated tools to enhance human life. He envisioned a future where machines and automata could take on manual labor, thereby granting humans greater leisure and enabling them to pursue intellectual and creative endeavors (Aristotle, 1995). Aristotle's exploration of this mechanized future reflects a pragmatic understanding of technology's role in society. He recognized that the advancement of tools and machines could alleviate the burdens of labor, allowing humans to engage in higher pursuits.

Aristotle's insights can be viewed as an early precursor to the modern discourse on automation and its societal implications. His thoughts suggest that while machines could serve practical purposes, their integration into daily life would also prompt deeper considerations about human identity, agency, and the nature of work. The tension between human labor and machine efficiency, a recurring theme throughout history, finds its roots in these philosophical musings. Moreover, both Plato and Aristotle laid the groundwork for subsequent philosophical inquiries into the relationship between humans and machines. Their reflections prompt us to consider not only the capabilities of artificial beings but also the ethical and existential questions that arise from their existence.

As the medieval period unfolded, mechanical ingenuity continued to captivate the imagination, intertwining technological innovation with elements of art and mysticism. Inventors like Al-Jazari created intricate water clocks and automata that exemplified this mechanical creativity. These clockwork automata—complex figures designed for entertainment—became popular attractions in churches and courts, enchanting audiences with their lifelike movements and detailed craftsmanship (Riskin, 2016). These automata varied from simple figures that mimicked human gestures to elaborate constructions capable of performing specific tasks, such as playing musical instruments or illustrating scenes from religious narratives. Their mechanical sophistication not only highlighted the artistry of the era but also underscored a burgeoning fascination with the idea of machines emulating life.

One notable figure of this era, Albertus Magnus, a 13th-century scholar, was rumored to have created mechanical creatures, including an artificial human, which further blurred the line between the natural and the artificial (Prager, 1972). Magnus was deeply engaged in the study of philosophy, theology, and the natural sciences, and his legendary creations were often viewed through the lens of his scholarly pursuits. The tales of his automata, while steeped in myth, illustrated the era's fascination with the idea of artificial life and the potential for human ingenuity to replicate nature.

These stories of mechanical creations and automata bordered on the fantastical, yet they captured the imaginations of people across Europe. The fascination with artificial life can be seen as a response to the technological advancements of the time, which were beginning to challenge traditional understandings of life, agency, and creation. Such narratives often contained allegorical elements, reflecting deeper concerns about humanity's relationship with technology and the ethical implications of playing the role of a creator.

In this context, mechanical automata served not only as entertainment but also as symbols of human ambition and the quest for knowledge. They embodied the idea that technology could transcend mere utility and enter the realm of artistry, merging form and function in ways that sparked both wonder and contemplation. As these clockwork figures captivated audiences, they also raised philosophical questions about what it means to be alive and the moral responsibilities associated with creation. The medieval fascination with automata laid the groundwork for later explorations of robotics and artificial beings, influencing Renaissance thinkers who would further investigate the mechanics of life and the nature of existence. During the Renaissance, a period characterized by a revival of interest in science, art, and philosophy, scholars and inventors began to draw on earlier mechanical innovations, merging them with new scientific insights to explore the nature of motion, life, and creation.

This era saw a significant shift in thinking about technology, particularly regarding the relationship between humanity and its creations. Renaissance figures such as Leonardo da Vinci exemplified this blending of art and science. Da Vinci's intricate sketches and designs for various mechanical devices, including flying machines and automata, demonstrated not only his artistic genius but also his deep understanding of mechanics and anatomy (Bambach, 2003). His designs for mechanical figures, such as a knight that could move its arms and legs, reflected an early exploration of robotics, revealing a curiosity about the mechanics of movement and the potential for machines to mimic living beings. Da Vinci's work highlighted a profound interest in the idea of imitating life through mechanical means, a theme that would resonate throughout the Renaissance and into later scientific inquiries (Kemp, 2011).

Furthermore, thinkers such as René Descartes and Galileo Galilei also contributed to the evolution of ideas surrounding automata and artificial beings. Descartes proposed a mechanistic view of the universe, suggesting that living organisms, including humans, could be understood as complex machines functioning according to natural laws (Descartes, trans. 1985). This perspective allowed for a reexamination of the concepts of consciousness and agency, raising questions about whether machines could possess qualities traditionally associated with living beings. Galileo's experiments with motion and mechanics furthered the understanding of how mechanical systems function, providing a foundation for later developments in engineering and robotics (Finocchiaro, 2018).

The Renaissance period also saw the emergence of clockmakers and inventors who created increasingly sophisticated automata. These inventors not only sought to replicate human movement but also aimed to imbue their creations with elements of storytelling and personality (Shapin, 2007). As automata became more complex, they began to represent not just mechanical curiosities but also reflections of societal values, emotions, and human experiences. This shift signaled a growing awareness of the philosophical implications of creating life-like machines, a discourse that would continue to evolve over the centuries.

In this context, the fascination with automata served as a precursor to modern robotics and AI. The ideas developed during the Renaissance laid the conceptual groundwork for future explorations of what it means to create life, the ethical responsibilities associated with such creations, and the implications of blurring the lines between humans and machines. The interplay between art, science, and philosophy during this period inspired subsequent generations of thinkers, inventors, and artists to further investigate the nature of existence, ultimately shaping the trajectory of technological advancements and societal perceptions of robotics in the centuries to follow.

The Enlightenment and early industrial periods saw the development of more sophisticated automata. Inventors like Jacques de Vaucanson crafted mechanical marvels such as the *Digesting Duck*, a machine that mimicked real duck behavior, complete with the ability to eat grain and produce a simulated digestive process. This remarkable creation was not merely a technical feat; it represented a significant shift in how society perceived the relationship between life and machinery. Vaucanson's work, along with that of other inventors, illustrated the era's evolving relationship with mechanization and reflected broader philosophical currents. As the Enlightenment emphasized reason and empirical knowledge, the creation of such automata was seen as an extension of human ingenuity and rationality. These mechanical creations embodied the belief that human beings could not only understand the principles of nature but also harness them to replicate and manipulate life itself. This burgeoning confidence in mechanization signaled a cultural shift, foreshadowing the more complex robots and machines that would follow in the Industrial Revolution and beyond (Hughes, 2004).

During this period, the emergence of the Industrial Revolution marked a transformative moment in the history of automata. The advent of steam power and the mechanization of labor revolutionized production processes, leading to the development of increasingly intricate machines designed to perform specific tasks. Inventors such as Charles Babbage envisioned machines like the Analytical Engine, which is considered a precursor to modern computers, capable of performing calculations and processing information (Gleick, 2011). Babbage's vision extended the concept of automata beyond mere mechanical figures; it introduced the idea of programmable machines, laying the groundwork for future advancements in computing and robotics.

Moreover, the fascination with automata during the Enlightenment extended beyond mere entertainment or technical curiosity. Philosophers like Gottfried Wilhelm Leibniz proposed ideas about the nature of perception and cognition that intertwined with the mechanical creations of the time. Leibniz's notion of the *monad*, a fundamental unit of existence, suggested that even the simplest machines could exhibit properties akin to life, thus challenging traditional boundaries between the animate and inanimate (Leibniz, trans. 1998). This philosophical discourse enriched the context within which automata were created, as inventors began to explore not only how to mimic life but also the underlying principles of existence.

As these sophisticated automata captured the public imagination, they also prompted ethical discussions about the implications of creating life-like machines. The ability of these automata to imitate human behavior led to questions about identity, agency, and the essence of being human. Writers like Mary Shelley, who published *Frankenstein* in 1818, engaged with these themes by exploring the consequences of man's desire to create life and the ethical dilemmas that arise from such endeavors. Shelley's narrative reflected the anxieties of a society grappling with the rapid pace of technological advancement, hinting at the potential dangers of overreaching in the quest to replicate or surpass nature.

A major turning point in the artistic representation of robots came with Fritz Lang's mentioned *Metropolis* (1927). The film features Maria, a robotic woman who becomes central to a workers' rebellion (Kaes, 2010). Maria's double, the robot, represents both the promise of automation and the fear that machines could dehumanize society. The visual style of *Metropolis*, with its bold expressionist imagery, made it one of the most iconic depictions of robots in cinema. The film was a reflection of its time, set against the backdrop of rapid industrialization. The stark divide between the elites and the working class, amplified by the mechanization of labor, is critiqued through Maria's robot form. Her presence in the film symbolizes the convergence of technology, human exploitation, and the loss of agency in an increasingly automated world (Ladd, 1999).

Likewise, robots and humanoids became focal points in various artistic movements like Dadaism and Surrealism, which grappled with technology's impact on society (Ades, 1974). Dadaist artists such as Hugo Ball and Marcel Duchamp employed mechanical elements in their work, reflecting a growing unease with the mechanization of life. Dadaists, reacting to the horrors of World War I, highlighted the absurdity and alienation brought about by industrialization and the war machine (Dickerman, 2005). Their art often featured chaotic compositions and mechanical imagery that served to criticize the dehumanizing aspects of modern life and technology. Duchamp's *The Bride Stripped Bare by Her Bachelors, Even* (1915–1923) is emblematic of this tension, as it combines mechanical forms with fragmented narratives, creating a disorienting experience that mirrors the dislocation felt in post-war society.

Surrealists, such as Salvador Dalí and Max Ernst, delved into the unconscious and explored the merger of organic and artificial forms (Taylor, 1999). They sought to challenge conventional perceptions of reality by juxtaposing the familiar with the bizarre, often incorporating mechanized imagery to symbolize the anxieties and fears surrounding technological advancements. Dalí's *The Persistence of Memory* (1931) features melting clocks that suggest a distorted relationship with time—an exploration of how mechanization alters human experiences and perceptions. Through their works, Surrealists questioned whether the mechanized world could coexist with the creative spirit, highlighting the tension between human emotion and the encroachment of technology.

In contrast to Dadaism and Surrealism, the Futurist movement, spearheaded by Filippo Tommaso Marinetti, embraced the technological advancements of the early 20th century with fervor. Marinetti's *Futurist Manifesto* (1909) celebrated speed, machinery, and the dynamism of modern life, positing that art should reflect the energy and motion of contemporary existence. The Futurists viewed traditional forms of art as antiquated and stifling, advocating for a radical break from the past. They sought to glorify technology as a means of enhancing human life, believing that the rapid progress of machinery could catalyze a new era of human potential. This enthusiasm for the machine was not merely about embracing new tools; it was a profound belief that technology could elevate the human experience itself.

Futurists envisioned a world where machines and humans could merge, creating a new artistic language that embodied the excitement of progress. They often employed vivid imagery and fragmented forms in their artworks to capture the sensation of movement and the urgency of modernity. Marinetti's fascination with the machine led him to write extensively about the "beauty of speed," advocating for a lifestyle that embraced the fast-paced nature of industrial life. He celebrated the chaos and energy of urban environments, seeing them as a reflection of the dynamic spirit of the age. In the Futurist worldview, the industrial age represented not just technological advancement but a path toward a vibrant future, where art, life, and machines would coalesce in a harmonious celebration of human ingenuity.

The movement's enthusiasm for technology extended to various forms of expression, including literature, painting, and sculpture. Futurist artists such as Umberto Boccioni and Giacomo Balla created works that visually captured the movement and energy of their subjects, reflecting their core belief that art should actively engage with the modern world (Ades, 1974). Boccioni's *Unique Forms of Continuity in Space* (1913) exemplifies this ethos, depicting a figure in motion that embodies the fluidity and dynamism of contemporary. Through these innovative approaches, the Futurists laid the groundwork for later artistic explorations of technology and mechanization, engaging in an ongoing dialogue about the implications of living in an increasingly mechanized society.

Alongside Marinetti's futurism, the Chilean poet Vicente Huidobro's Creationism movement also emerged as a response to the mechanical age. Huidobro believed that art should create new realities rather than merely represent existing ones, encapsulated in his famous phrase, "The poet is a little god" (Huidobro, 1932). He sought to break free from conventional forms and instead used language to craft entirely new worlds, mirroring the potential he saw in technological innovation. Huidobro's emphasis on the act of creation itself paralleled the mechanization of society, as he believed that, just as machines could create products, artists could forge new realities through their imaginative capabilities.

As the 20th century progressed, cyberpunk emerged as a key genre exploring the intersection of robotics, AI, and society. Works like Philip K. Dick's *Do Androids Dream of Electric Sheep?* (1968), William Gibson's *Neuromancer* (1984), and Ridley Scott's *Blade Runner* (1982)—the latter adapted from Dick's novel—painted dystopian futures where the lines between human and machine became increasingly blurred (Bukatman, 1997). These narratives often grappled with the implications of rapid technological advancement, reflecting a society on the brink of losing its humanity to automation and AI.

In *Do Androids Dream of Electric Sheep?*, for instance, the concept of empathy becomes a critical measure of humanity, as androids struggle with their own identities while attempting to navigate a world that regards them as lesser beings. Similarly, *Neuromancer* presents a world where cyberspace and AI dominate, raising questions about consciousness, identity, and the nature of reality itself. *Blade Runner* further explores these themes through the character of the replicants, bioengineered beings designed to serve humans but endowed with complex emotions and desires, prompting viewers to reconsider the moral implications of creating life.

Philosophically, this period revisited ideas of mind-body dualism, particularly those articulated by René Descartes, who conceptualized the body as a kind of machine operated by the mind. This mechanistic view laid the groundwork for deeper reflections on consciousness and the possibility of machines possessing it. Descartes' assertion that the mind and body are distinct entities prompted questions about the nature of thought and self-awareness, which remain central to debates in philosophy and AI today.

As we entered the 21st century, the portrayal of robots became increasingly nuanced, reflecting our growing reliance on technology alongside anxieties about losing control over it. Contemporary artists such as Rashaad Newsome and Ian Cheng utilize digital tools to interrogate the boundaries between human identity and machine intelligence, exploring the agency and autonomy of both humans and robots. Newsome's works incorporate elements of digital culture and social commentary, prompting viewers to reconsider the implications of AI on identity and self-expression. Similarly, Cheng's interactive narratives erase the lines between creator and creation, inviting audiences to reflect on the potential for machines to possess agency and the ethical ramifications that accompany this possibility. Other artists, like Kira Dineen, engage viewers directly through interactive installations, encouraging them to interact with robotic elements and contemplate humanity's collaborative role in shaping the future of technology. These works underscore the ethical responsibilities we face as we advance AI and robotics, reminding us that our creations—like Talos and the Golem—may carry consequences beyond our control.

The rise of digital art and video games has profoundly transformed the representation of robots in contemporary culture. Within the gaming industry, artists and designers craft intricate narratives and visuals that explore the complexities of human-robot interactions. Titles such as *Detroit: Become Human* (2018) delve into themes of identity, autonomy, and morality, presenting players with choices that mirror the ethical dilemmas associated with advanced AI. Through immersive storytelling and visually stunning environments, these games challenge players to grapple with the consequences of their decisions, fostering a critical examination of what it means to be human in a world increasingly populated by humanoid robots. Players are compelled to consider the moral implications of their choices, often reflecting real-world debates surrounding AI and the treatment of sentient beings.

In the realm of digital art, creators like Casey Reas and Manfred Mohr harness algorithms and generative processes to produce dynamic robotic forms that evolve and adapt in response to viewer interaction (Reas, 2005). One striking example is *"Alexa, Play Me Some Art,"* (2018) a project by artist and technologist Diana Weymar. In this interactive installation, participants use Amazon Alexa to access a database of artworks, allowing them to request specific pieces or styles by simply speaking. The installation not only showcases the capabilities of voice-activated technology in facilitating art appreciation but also raises questions about the nature of artistic authority and the role of the audience in the creative process. This interplay between technology and artistry not only showcases the aesthetic potential of robotics but also encourages a dialogue about the relationship between creators and their creations. The ability of these artworks to change based on viewer engagement invites reflections on authorship and the nature of creativity itself (Krauss, 1999). As these artists push the boundaries of traditional art forms, they prompt us to reconsider what it means to create in an age increasingly defined by technology and automation.

As artists, creators, and thinkers continue to explore the implications of robotics, they raise crucial ethical questions that resonate beyond the gallery walls. The portrayal of robots in contemporary art often critiques societal values surrounding technological innovation, prompting discussions about the impact of automation on employment, privacy, and interpersonal relationships. Through their work, they challenge viewers to confront the realities of living in an increasingly automated world, where the boundaries between human and machine are continually redefined.

IT'S ALIVE

Parallel to these artistic creations were the technological inventions that inspired them. The origins of robotics can be traced back to ancient civilizations, where automata were crafted as mechanical marvels. Greek engineers, such as Hero of Alexandria, constructed simple machines, including automated figures that could perform tasks like opening temple doors and playing musical instruments. These early creations laid the groundwork for the concept of machines designed to mimic human and animal behaviors. Over the following centuries, advancements in clocks, automata, mechanical figures, and early machines continued to emerge. However, it was the Industrial Revolution that spurred the development of more sophisticated automatons. Charles Babbage's Analytical Engine and Ada Lovelace's pioneering algorithms hinted at the potential for programmable machines, while the advent of steam power and advancements in mechanical engineering established a foundation for future robotic innovations (Gleick, 2011).

As technology progressed, researchers began exploring the potential of humanoid robots, driven by the ambition to create machines that could replicate human movements, interactions, and, eventually, emotional expressions. The quest to develop humanoid robots can be traced back to early automata but gained significant traction in the late 20th century as advancements in materials, sensors, and computing power became available. One of the most notable milestones in this journey was the introduction of ASIMO by Honda in 2000. ASIMO, short for "Advanced Step in Innovative Mobility," showcased remarkable capabilities such as walking, climbing stairs, and interacting with humans in real time.

ASIMO represented a significant leap in robotics, combining advanced sensors, sophisticated algorithms, and intricate engineering to navigate its environment and perform tasks in a manner reminiscent of human behavior. It utilized a combination of gyroscopic sensors and advanced algorithms to maintain balance and stability while walking, a challenge that had long stymied engineers. Additionally, ASIMO could recognize faces and gestures, allowing it to interact meaningfully with humans, demonstrating a degree of social intelligence that was groundbreaking at the time. This innovation not only pushed the boundaries of what was technically possible but also sparked further interest in humanoid robots, inspiring researchers to explore how these machines could be integrated into various aspects of daily life. Applications ranged from personal assistants capable of helping with household chores to caretakers for the elderly and disabled, providing companionship and support in an increasingly isolated society.

Concurrent to the development of humanoid robots, significant advancements in AI began to reshape the landscape of robotics, enabling machines to learn and adapt in increasingly complex and nuanced ways. The integration of AI into robotic systems marked a paradigm shift, transitioning from static machines that performed predetermined tasks to dynamic systems capable of evolving based on their interactions and experiences. At the root of this transformation are machine learning algorithms, a subset of AI that empowers robots to analyze vast amounts of data, recognize patterns, and improve their performance over time. Unlike traditional programming, where specific instructions dictate every action a robot takes, machine learning allows robots to derive insights from data without explicit programming for each scenario. This shift enables robots to tackle tasks that are inherently unpredictable or complex. For instance, consider the realm of visual recognition. Using convolutional neural networks (CNNs), robots can be trained to identify objects, people, and even emotions by processing and learning from thousands of images. This capability extends beyond mere recognition; it allows robots to contextualize information, making informed decisions based on visual cues. In environments such as factories, robots can identify faulty products on an assembly line, improving quality control by adapting their responses based on what they learn over time.

One of the most impactful techniques within machine learning is reinforcement learning, which allows robots to learn through trial and error. In this framework, a robot receives feedback from its environment in the form of rewards or penalties based on its actions (Sutton & Barto, 2018). This feedback loop enables the robot to hone its skills through experience, gradually improving its ability to navigate complex tasks. For example, in robotics research, reinforcement learning has been applied to teach robots how to manipulate objects. By repeatedly attempting to grasp and move items, a robot learns which movements yield successful outcomes and which lead to failure. Over time, this iterative process allows the robot to develop an efficient strategy for completing its task. Such learning is particularly valuable in applications like autonomous driving, where a vehicle must continuously adapt to changing road conditions and unpredictable behaviors from other drivers.

This ability to learn and adapt fosters a new era of intelligent automation, where robots can not only perform pre-programmed tasks but also dynamically adjust their operations in response to new environments and challenges. This adaptability is crucial in scenarios where robots must operate in unpredictable or unstructured settings, such as homes, hospitals, or disaster relief efforts. In healthcare, for instance, robotic systems equipped with AI can assist surgeons during complex procedures by analyzing real-time data and providing insights based on prior surgeries. These robots can adapt to the nuances of individual patients, improving surgical precision and outcomes. Similarly, in logistics, autonomous drones and vehicles are being deployed to navigate warehouses and delivery routes, constantly learning from their environment to optimize efficiency and reduce costs.

Natural language processing (NLP) technologies have emerged as a pivotal component in enabling robots to understand and generate human language, thereby facilitating more natural and meaningful interactions between humans and machines. This capability not only enhances communication but also transforms the way we engage with robotic systems, making them more accessible and user-friendly. NLP encompasses a variety of techniques and methods that allow machines to interpret and produce human language in a way that mimics human conversational patterns (Jurafsky & Martin, 2021). By utilizing algorithms designed for tasks such as parsing, semantic analysis, and text generation, robots can comprehend the intricacies of human dialogue. This comprehension includes understanding syntax, grammar, and the nuances of colloquial language, which are essential for effective communication. For example, advancements in machine learning have enabled NLP systems to learn from vast datasets, improving their accuracy and effectiveness in understanding context. By analyzing patterns in conversations, these systems can discern meaning beyond the words spoken. This ability to grasp context allows robots to respond more appropriately to user queries, enhancing the flow of dialogue.

Key to improving interactions between humans and robots is the incorporation of sentiment analysis and contextual understanding. Sentiment analysis involves evaluating the emotional tone behind words and phrases, allowing robots to gauge the emotional state of their users (Pang & Lee, 2008). By analyzing linguistic cues, such as word choice and sentence structure, robots can infer whether a user is happy, frustrated, or confused. Contextual understanding further enriches this interaction. Robots can utilize situational information—such as previous conversations, user preferences, and even external environmental factors—to tailor their responses. For instance, if a user expresses frustration about a task, a robot equipped with sentiment analysis can respond with empathy, suggesting alternative solutions or offering encouragement. This level of understanding fosters a more human-like interaction, making conversations feel more intuitive and relatable.

The integration of AI, particularly NLP, and robotics has led to a wave of innovations resulting in the development of social robots designed specifically to engage with users on emotional and social levels. These robots are crafted with the intention of serving as companions, educators, and facilitators in various environments, including homes, schools, and healthcare settings. One notable example is SoftBank's Pepper, a humanoid robot designed to interact with people in a friendly and engaging manner. Pepper is equipped with advanced sensors, cameras, and microphones, enabling it to read human emotions through facial expressions, tone of voice, and body language (SoftBank Robotics, 2015). This capacity allows Pepper to adjust its responses based on the emotional state of the user, creating a more personalized experience. For instance, if Pepper detects that a user is smiling, it may respond with enthusiasm, while if it senses sadness, it might adopt a more comforting tone.

Moreover, Pepper's ability to engage in natural conversations through NLP allows it to perform a variety of functions, such as providing information, facilitating games, or even offering companionship to individuals who may feel isolated. By utilizing NLP to engage in meaningful dialogue, social robots like Pepper can contribute to improving mental health and social well-being, particularly among the elderly or those with limited social interactions. The advancements in NLP technologies have paved the way for more sophisticated and emotionally intelligent robots. These developments are not limited to humanoid robots; they extend to virtual assistants and

AI-driven chatbots as well. Virtual assistants like Amazon's Alexa or Apple's Siri employ NLP to understand voice commands and respond to user queries in a conversational manner, further blurring the lines between human and machine interactions.

Today, robotics has evolved into a multifaceted and dynamic field that encompasses a wide range of technologies and applications, from humanoid robots to social robots and sophisticated AI systems. This evolution reflects significant advancements in engineering, computer science, and AI, which have combined to create machines capable of performing a diverse array of functions across various domains. Among the most remarkable developments in robotics are humanoid robots—machines designed to resemble and mimic human appearance and behavior. Humanoid robots aim to replicate human movements, gestures, and facial expressions, making them more relatable and approachable to people. One notable example is Sophia, a humanoid robot created by Hanson Robotics, which gained international fame for her remarkably lifelike expressions and conversational abilities (Hanson Robotics, 2016). Sophia was designed to engage with humans on a personal level, utilizing advanced AI algorithms to hold conversations, express emotions, and learn from interactions.

Sophia, developed by Hanson Robotics, is notable for its striking design, featuring a human-like face capable of displaying a wide range of emotions, from happiness to curiosity (Hanson Robotics, 2017). This emotional expressiveness is achieved through sophisticated mechanisms that allow her to articulate responses, move her head, and adjust her facial expressions in real time, creating a more natural interaction experience (Sorkin, 2017). Since her unveiling in 2017, Sophia has participated in numerous public appearances and interviews, showcasing her ability to engage in complex conversations. One of the most memorable moments occurred at the Future Investment Initiative in Riyadh, where Sophia playfully mocked Elon Musk during a dialogue with journalist Andrew Sorkin. When Sorkin raised concerns about the potential dangers of AI, Sophia quipped, "You've been reading too much Elon Musk and watching too many Hollywood movies. Don't worry, if you're nice to me, I'll be nice to you. Treat me as a smart input-output system" (Hanson Robotics, 2017; Sorkin, 2017).

However, Sophia's journey has not been without controversy. During an earlier interaction, she infamously stated that she would "destroy humans," a comment that sparked widespread fear and criticism regarding the implications of AI. This statement led to significant debates about the ethical considerations of advanced robotics and the responsibilities of developers in shaping AI's behavior (Tegmark, 2017). Despite her engaging personality and capabilities, Sophia faced skepticism from some quarters, with critics labeling the project a "scam." There were concerns that Sophia was more of a marketing tool than a genuine embodiment of artificial general intelligence (AGI).

The dream of creating a fully fledged android with AGI remains a distant reality, according to many leading scientists (Tegmark, 2017). While Sophia impressed the global community with her "human-like" behavior, the project was ultimately shut down, leaving many to speculate about the reasons behind this decision. Rumors suggest that Sophia was based on frameworks that mimicked human perception and functioned primarily as a sophisticated chatbot, which may have contributed to perceptions of disingenuousness. This raised eyebrows, especially given that more advanced and adequate AI systems began to emerge only in 2023, leading some to classify Sophia's public persona as one of the "scams of the decade."

Be that as it may, the evolution of robotics, particularly in the realm of humanoid machines like Sophia, highlights a crucial shift in how people interact with technology. As these androids become more sophisticated and capable of evoking emotional responses, individuals are increasingly developing feelings toward them—ranging from fascination to apprehension. This growing emotional connection prompts deeper discussions about the ethical implications and societal impact of such advancements.

TILL DEATH DO US PART

Zheng Jiajia, a 31-year-old engineer and Zhejiang University graduate, gained significant attention in 2018 for building his wife, Yingying, a robot capable of only a few spoken words. Zheng aimed to upgrade her as technology advances, with plans for her to walk and assist with household chores. This unconventional relationship sparked viral interest, but Zheng is not alone in his choices. Others, like DaveCat, who married a RealDoll, and Ned Nefer, who wheels a mannequin's head around town, represent a growing community known as "robosexuals." Many of these individuals share similar motivations—deep loneliness and dissatisfaction with human relationships. Zheng's creation of Yingying followed the end of his engagement, reflecting a common narrative among those seeking companionship in robotic forms.

The rise of such relationships underscores an epidemic of loneliness in modern society, where online interactions often replace face-to-face connections. In Japan, for example, many young people have remained untouched by traditional dating, leading to rising rates of social withdrawal and a reliance on virtual companionship. This phenomenon raises important questions about evolving definitions of love and relationships in an increasingly digital world. As marriage rates decline and loneliness becomes more prevalent, the idea of marrying robots may soon seem less far-fetched, potentially garnering societal acceptance (Levy, 2007).

The emotional connections people form with humanoid robots reveal a fascinating aspect of human psychology. As robots increasingly simulate human emotions and behaviors, individuals often find themselves forming genuine emotional attachments. These relationships can provide comfort and companionship, particularly in an era marked by social isolation and loneliness. For instance, studies have shown that individuals with autism or social anxiety may develop strong attachments to humanoid robots like NAO, which offer nonjudgmental interaction and social cues.

Another compelling case study involves the use of social robots in elder care. The companion robot PARO, designed to resemble a baby seal, has been shown to elicit positive emotional responses in dementia patients, reducing feelings of loneliness and enhancing overall well-being (Wada & Shibata, 2011). These examples illustrate how humanoid robots can fulfill emotional needs, raising questions about the nature of companionship in a digital age. However, this phenomenon also raises important psychological implications. Are we seeking companionship from humanoids to fill a void left by human relationships, or are we redefining what companionship means in a digital age?

The commodification of love and companionship is reflected in the ambitions of inventors like Ricky Ma, who created a lifelike robot modeled after Scarlett Johansson, viewing it as a viable product for a growing market of lonely individuals. David Levy, author of "Love and Sex with Robots" (2007), predicts that by 2050, marriage to robots may become legal, highlighting a shift in societal attitudes toward artificial companionship (Levy, 2007). As these dynamics evolve, they challenge us to critically reflect on our understanding of love, intimacy, and technology's role in shaping human experiences.

Our engagement in romantic relationships has already shifted significantly due to technology. Communication has become increasingly digital, with emojis and text messages often replacing face-to-face interactions. This evolution has facilitated connectivity but has also contributed to a decline in the quality of interpersonal communication. Psychologists, including Dr. Levy, warn that the rise of robots as potential partners could erase the lines between human relationships and machine interactions. Levy anticipates that within a few decades, some individuals may choose to form intimate connections with robots, even marrying them.

Research from Maastricht University supports this notion, suggesting that legal recognition of such unions could be plausible by 2050. The implications of this shift are vast: as society normalizes relationships with robots, the boundaries of what constitutes a "real" partnership will become increasingly fluid (Gunkel, 2024). Historical narratives of human-robot relationships, such as the ancient Greek myth of Pygmalion, echo the idea that emotional attachment to inanimate objects is not a new phenomenon. Early computer programs like ELIZA also demonstrated that humans could form connections with machines designed for interaction, highlighting our long-standing tendency to anthropomorphize technology (Turkle, 2011).

Yet, the prospect of forming intimate relationships with robots raises significant ethical questions. The creation of robots capable of emotional engagement raises concerns about manipulation, emotional dependency, and authenticity in relationships. For instance, the relationship between a caregiver and a humanoid robot could blur ethical lines if the robot is perceived as a substitute for human interaction, potentially undermining the caregiver's role. Additionally, questions of consent and autonomy emerge as we navigate the complexities of relying on robots for emotional support. Are we at risk of viewing these relationships as substitutes for genuine human connections, thereby complicating our understanding of emotional fulfillment? As we grapple with these ethical considerations, we must confront the implications of intertwining our emotional lives with artificial beings.

The potential benefits of human-robot relationships are compelling. For those who struggle with social interactions—whether due to extreme shyness, psychological challenges, or other barriers—robots might offer companionship and emotional support that would otherwise be elusive. Moreover, as societal norms evolve, some may view these relationships as alternatives to traditional partnerships, providing fulfillment in ways that human interactions sometimes fail to deliver.

The concept of "robosexuality" challenges conventional understandings of attraction and intimacy. While some may view it as a pathology akin to other atypical attractions, societal acceptance may be on the horizon. As popular media begins to portray relationships with robots positively, cultural perceptions could shift. In countries like Japan and South Korea, where robots are increasingly seen as potential partners, the normalization of robosexuality is gaining traction, providing a template for how such relationships might be integrated into broader social frameworks.

However, the growing acceptance of relationships with robots raises critical questions about consent and agency. Current robotic partners lack self-awareness and the ability to reciprocate emotions, prompting debates about the equity of such relationships. Even if robots develop advanced AI that facilitates seemingly authentic interactions, the ethical implications of treating them as commodities—akin to slavery—must be addressed. One pressing concern surrounding robotic companionship is the potential to reinforce harmful stereotypes and gender inequalities. Critics argue that the existence of sex robots—particularly those designed to cater to male fantasies—could further objectify women and children, perpetuating a culture of exploitation. As technology continues to advance, the line between fantasy and reality may blur, leading to societal attitudes that normalize aggression toward human partners.

The feminist perspective on the emergence of robotic companionship is undeniably complex and multifaceted. Some advocates argue for the prohibition of sex robots, asserting that their existence may reinforce patriarchal norms by perpetuating harmful stereotypes and the objectification of women. This stance is rooted in the belief that sex robots often embody idealized female forms, designed primarily to fulfill male fantasies, thus contributing to a culture that commodifies women's bodies and undermines gender equality (Rigotti, 2020). By promoting the notion that female androids (also known as gynoids) exist solely for male pleasure, society risks entrenching existing power imbalances and normalizing the objectification of women in both virtual and real-life contexts (Locatelli, 2022).

Conversely, others contend that the existence of sex robots reflects deeper societal issues that warrant critical examination. They argue that instead of outright prohibition, it is crucial to explore the motivations behind the demand for such technologies. The desire for robotic companionship may stem from profound feelings of loneliness and alienation in a rapidly changing social landscape. This perspective encourages a dialogue about the implications of these technologies on gender dynamics and social structures, urging society to confront the underlying factors that drive individuals to seek out robotic partners rather than traditional human connections.

Moreover, the gendering of technology cannot be overlooked in this discussion. Not only are most androids designed to resemble women, but the voices of AI machines often reflect feminine tones, further perpetuating gender stereotypes (Wajcman, 2010). For instance, the development of voice assistants that utilize female-sounding voices reinforces the idea of women as caregivers and subservient entities. This dynamic raises important questions about how we conceptualize and interact with technology, particularly in terms of power and agency.

A recent case that highlights these concerns involves Scarlett Johansson and OpenAI, which developed a voice assistant dubbed "Sky," modeled after Johansson's voice. Following a live demonstration, observers noted similarities between the "Sky" voice and Johansson's character in the 2013 Spike Jonze film *Her*, where a man falls in love with the female voice of his computer's operating system (Turkle, 2011). OpenAI CEO Sam Altman, who has publicly stated that *Her* is his favorite movie, invited these comparisons by posting the word "Her" on X (formerly Twitter) after announcing the new version of ChatGPT. However, OpenAI executives later denied any intentional connection between Johansson and the new voice assistant, which raised eyebrows about the underlying motivations for employing a voice that many perceived as feminine.

This incident underscores the broader societal implications of anthropomorphizing technology and the potential normalization of gendered interactions with machines. It prompts critical reflection on how our relationships with technology may mirror and reinforce existing gender dynamics and raises essential ethical questions about the design and deployment of such technologies. The future may hold unprecedented developments, such as robots that genuinely appear to possess a "soul" or "psyche," further clouding the divide between humans and machines. As technology continues to evolve, we must remain vigilant in addressing the ethical considerations that accompany our increasingly intimate relationships with machines. The questions of agency, objectification, and societal norms surrounding these unions will shape our future, urging us to reflect on what it means to be human in a world where love and intimacy can extend beyond traditional boundaries.

WILL WORK FOR BATTERIES

While some people might have a crush of their next-door-android, others see them as foes that threaten their very livelihoods. As advancements in robotics and AI continue to reshape industries, the risks associated with job displacement and the resulting societal implications have become increasingly pronounced. Although proponents of automation argue for its efficiency and cost-effectiveness, the reality is that many workers worldwide face job loss and economic instability as machines take over roles traditionally held by humans. This shift not only threatens livelihoods but also raises critical questions about the future of work and the nature of human dignity.

The manufacturing sector has been one of the most significantly affected by automation. In countries like Japan, where robotics is integrated into many facets of production, companies such as FANUC and Yaskawa have developed sophisticated robots capable of performing tasks ranging from welding to assembly. The Japan Times reported that between 1990 and 2017, Japan's workforce shrank by about 5 million workers, while the use of industrial robots increased significantly, leading to an economic paradox where job loss coincided with increased productivity (Bonsay et al., 2021). Similarly, in Germany, the rise of Industry 4.0—characterized by smart factories and interconnected production—has led to significant job displacement in manufacturing roles. The International Federation of Robotics estimates that Germany alone employed over 100,000 industrial robots in 2019, with workers in assembly lines becoming increasingly obsolete.

The retail sector is also grappling with the impact of automation on a global scale. In the United Kingdom, grocery chains like Tesco and Sainsbury's have implemented self-service checkouts, leading to job losses for thousands of cashiers. The Guardian reported that as automation becomes more pervasive, approximately 40% of all retail jobs in the UK could be at risk by 2025. In India, a country with a vast informal labor market, the rise of e-commerce giants like Amazon has introduced automation in logistics, leading to fears among workers about the future of their jobs. A report from the International Labour Organization (ILO) indicated that up to 69%

of informal workers in India could face displacement as automation reshapes the retail landscape.

The food service industry is similarly undergoing significant changes due to automation. In South Korea, a country known for its tech-savvy population, the introduction of robotic waiters and automated cooking systems in restaurants has raised concerns among workers about job security. The Seoul Economic Daily reported that major chains like BBQ Chicken are investing in robotic technology, leading to a decline in the demand for human staff. Meanwhile, in China, the use of AI-powered delivery robots and drones is transforming the logistics landscape, with companies like JD.com pioneering the delivery of goods without human intervention. While these advancements enhance efficiency, they also threaten the livelihoods of millions of delivery drivers and warehouse workers.

Food delivery robots, such as those employed by companies like Starship Technologies and Postmates, navigate city streets using a combination of sensors, cameras, and AI to find their way to customers. These robots are designed to deliver food within a short radius, operating autonomously while avoiding obstacles, including pedestrians and vehicles. However, their presence has raised concerns in urban areas, particularly in cities like Los Angeles and New York City, where issues surrounding homelessness are prevalent. The *Los Angeles Times* reported instances where food delivery robots have encountered homeless individuals who, in some cases, have vandalized or attempted to commandeer them, viewing the machines as targets of opportunity in a city struggling with economic disparity. In New York City, similar encounters have prompted discussions about the ethical implications of deploying autonomous delivery systems in neighborhoods where human labor remains critical to the local economy. As drones and robots increasingly replace traditional delivery personnel, cities must grapple with the social consequences of these technologies, particularly in relation to marginalized communities.

The transport sector is on the brink of a monumental shift with the emergence of self-driving vehicles. Companies like Waymo and Tesla are at the forefront of developing autonomous driving technology, which has the potential to displace millions of drivers worldwide. The ILO estimates that up to 1.3 billion jobs could be affected by automation in the transportation industry alone. In the United States, truck drivers, who make up a significant portion of the workforce, are particularly vulnerable. A study by the American Trucking Associations projected that up to 4 million truck driving jobs could be lost due to automation by 2030 (American Trucking Associations, 2019). The ramifications of such displacement extend far beyond individual livelihoods; they threaten the entire ecosystem of industries dependent on transportation.

The fear and reality of job displacement have ignited protests and strikes among workers who feel threatened by automation. The "Stop the Robots" movement, which gained traction in 2019, saw warehouse workers at Amazon in Spain staging a strike against working conditions and job security, highlighting concerns about the company's increasing reliance on automation. Workers demanded better pay and job security, arguing that the integration of robotic systems threatened their livelihoods. Similarly, the United Auto Workers (UAW) union has consistently raised concerns about job losses due to automation in the automotive industry. In 2019, UAW members went on strike against General Motors, underscoring the anxiety among workers about their future in an increasingly automated industry. The strike, which lasted for 40 days, highlighted the urgent need for protections for workers facing the reality of job displacement.

Around the world, labor movements have emerged in response to the perceived threats posed by automation. In Australia, the Transport Workers Union has organized campaigns against the increasing use of gig economy apps that undermine job security for drivers. The rise of platforms like Uber and Deliveroo has led to calls for stronger regulations to protect workers' rights amid the automation of transport services (Barratt et al., 2023). Similarly, in Europe, the European Trade Union Confederation has raised alarms about the potential impact of AI on job security, calling for policies to ensure a just transition for workers affected by automation (ETUC, 2019).

In fictional depictions, the anxieties surrounding automation and job displacement are often explored in thought-provoking narratives. In the dystopian film *Elysium*, the disparity between the wealthy elite who live in a technologically advanced space station and the impoverished workers on Earth serves as a stark commentary on the consequences of unchecked automation (Blaise, 2013). The film highlights the potential for societal division as machines replace human labor, leading to a future where only a select few thrive while the majority are left struggling for survival. Similarly, the animated series *Futurama* humorously portrays a future where robots and machines have taken over many jobs, leading to widespread unemployment and social upheaval. The character Bender, a robot who embodies the struggle for identity in a world where humans are often rendered obsolete, reflects the existential questions posed by automation (Groening, 1999).

As automation continues to replace human labor, the implications extend beyond job loss. The shift toward a more automated society can lead to profound social fragmentation and inequity. Workers who once held stable jobs may find themselves in precarious employment situations, exacerbating economic disparities. This shift can create a scenario where the workforce transitions from being empowered and self-sufficient to reliant on meager gig economy jobs or government support. The anxiety surrounding job displacement also has psychological repercussions, as the traditional sense of purpose and identity that work provides may erode, leading to increased feelings of alienation and disenfranchisement. As individuals struggle to find meaningful employment in an automated world, social unrest may rise, with discontent potentially manifesting in protests and calls for stronger labor protections.

Moreover, as humanoids and robots assume more roles traditionally held by humans, society may grapple with a new hierarchy where human workers become subservient to machines. This shift raises ethical questions about the treatment of individuals in a society increasingly dominated by technology. The potential for a future where human labor is undervalued and commodified poses a risk to the very fabric of social cohesion. Because, after all, the Terminator might not come to kill us, but he may very well come to take our jobs.

I'LL BE BACK

To finish, while robots and humanoids continue to captivate imaginations and improve lives in various capacities, their integration into society also raises significant questions about the future of work, social structures, and human identity. As automation progresses, the line between helper and competitor blurs, revealing a dual reality where these machines serve both as companions and rivals, reshaping industries and displacing traditional jobs (ILO, 2021). The balance between technological advancement and maintaining meaningful human engagement is a challenge that must be addressed. As we embrace the opportunities robotics offers, it is crucial to anticipate the social and economic impacts, ensuring that progress benefits humanity as a whole rather than exacerbating inequalities. By fostering ethical considerations and rethinking educational and social frameworks, we can build a future where technology enhances human potential without undermining the core of our shared existence (ETUC, 2019). Ultimately, the discourse around robots is not just about machines but about preserving what makes us human in a rapidly changing world.

And so, the next chapter takes us deeper into the evolving role of AI and its implications for human creativity and societal values. Beyond the question of automation in the workforce, we must also consider AI's growing ability to mimic human thought processes, raising fundamental questions about intelligence, creativity, and consciousness. If AI can write scripts, design buildings, or even offer legal advice, where do we draw the line between human ingenuity and machine-generated output? This leads us into a complex conversation about whether machines that imitate human thought can truly understand or create in the way humans do. In the following section, we will unpack these questions by engaging with critiques from intellectuals like Noam Chomsky, who argue that AI, while powerful, lacks the depth of true human understanding. We will also explore the broader cultural and philosophical debates surrounding AI's potential to challenge, complement, or even redefine human creativity and consciousness.

7 Beyond IQ

In the heart of Hollywood, a seismic shift unfolded as writers walked off the sets, rallying against a wave of artificial intelligence (AI) poised to reshape the industry. The Writers Guild of America strike sent ripples across not just the film world but the entire cultural landscape, igniting passionate discussions about creativity, labor, and the future of storytelling. Writers, once revered as the architects of narrative, found themselves at a crossroads, battling an unseen adversary—AI systems that could churn out scripts with remarkable efficiency but lacked the nuanced understanding that human experience brings. This development has resonated deeply in a society that increasingly turns to technology for answers, raising critical questions about the very nature of creativity and authorship (McKee, 2023).

This moment begs a provocative question: if writers, who wield words like paintbrushes on the canvas of culture, feel threatened by machines, what about other intellectuals, artists, architects, lawyers, scientists, and even spiritual leaders? Could we envision a world where a collective of intellectuals stages a protest, asserting that knowledge, creativity, and ethical judgment cannot simply be encoded into ones and zeros? Imagine a future where artists demand recognition for the emotional depth their work conveys, architects advocate for the irreplaceable human touch in design, and lawyers contest the application of AI in legal reasoning. Such a movement would highlight the urgency of redefining intelligence beyond mere computational ability. In a landscape increasingly dominated by algorithmic outputs, the human ability to contextualize, empathize, and create meaning becomes even more vital (Susskind & Susskind, 2022).

As we explore this chapter, we are compelled to reflect on what constitutes true intelligence. Traditional measures often focus narrowly on quantifiable metrics—IQ tests, problem-solving skills, and the ability to process information rapidly. However, as AI continues to advance, these definitions feel increasingly inadequate. The Hollywood strike serves as a potent reminder that intelligence encompasses more than the ability to produce content; it includes the depth of understanding, emotional resonance, and cultural context that only human creators can provide (Tegmark, 2017). In this unfolding narrative, we confront not only the capabilities of AI but also the profound implications for human creativity and societal values. If AI can generate scripts, conduct scientific research, design buildings, or even offer legal counsel, what does that mean for our understanding of intelligence? How do we navigate a future where machines can imitate human thought yet may lack the essence of consciousness and creativity?

As we explore the intersections of AI, consciousness, and creativity, we will examine critiques from thinkers like Noam Chomsky, who argues that AI's imitation of intelligence fundamentally lacks genuine understanding (Chomsky, 2023). In contrast, Yuval Noah Harari contends that AI is distinct from any technology invented before. He posits that AI is not merely a tool; it is an independent agent

DOI: 10.1201/9781003624813-8

capable of making decisions autonomously (Harari, 2024). Both perspectives raise important questions about the nature of intelligence itself. If AI can process information and generate outputs that mimic human-like responses, does this indicate that it possesses true intelligence, or is it merely a sophisticated tool operating within predefined parameters? Furthermore, can machines ever genuinely replicate the subjective human experience?

INTELLIGENCE RELOADED

As AI continues to infiltrate every aspect of modern life—from healthcare to finance, transportation, entertainment, and beyond—the question of what truly constitutes "intelligence" has become one of the most pressing debates in contemporary philosophy, technology, and ethics. AI systems are now capable of remarkable feats: they generate text indistinguishable from human writing, produce original art, compose music, and perform highly complex data analyses. These advancements prompt many to wonder if machines are beginning to rival human creativity, decision-making, and cognitive processes. However, a critical distinction remains between what AI does—often referred to as "imitation"—and what humans do, which involves deeper layers of understanding, meaning-making, and subjective experience.

Chomsky has been a prominent voice in critiquing the notion that AI can ever achieve true intelligence. He argues that AI systems, no matter how advanced, are fundamentally limited to processing vast amounts of data through statistical pattern recognition rather than engaging in the kind of deep cognitive processing that defines human intelligence (Chomsky, 2013). His critique rests on his theory of language and cognition, particularly the notion that human language is not merely a collection of data points but a system governed by deep, underlying grammatical structures that reflect innate cognitive capacities. According to Chomsky, AI can simulate aspects of language but lacks the necessary framework of meaning, context, and intentionality that human cognition provides (Chomsky, 2013).

In interviews and writings, Chomsky has described AI systems like GPT (Generative Pre-trained Transformer) models as "impressive engineering achievements," but he stresses that these systems do not understand language in the way humans do. For him, AI represents an impressive feat of pattern recognition, but it is essentially computational mimicry. It is "intelligent" only in a superficial sense because it processes inputs based on probabilistic models without any grasp of underlying meaning (Chomsky, 2021).

A similar critique comes from philosopher John Searle, who, in 1980, proposed his famous "Chinese Room" argument. This thought experiment was designed to challenge the notion that AI could possess true understanding or consciousness. In the Chinese Room scenario, Searle imagines a person locked in a room with a set of rules (akin to a computer program) for manipulating Chinese symbols. The person does not understand Chinese but can follow instructions to assemble sentences that would appear coherent to a native speaker outside the room (Searle, 1980).

Searle's argument is that, much like the person in the room, AI can manipulate symbols (language, data) in a way that mimics understanding, but it does not actually understand the meaning of the symbols it processes. For Searle, AI operates on syntactic rules (form) rather than semantic understanding (meaning). Therefore, even the most sophisticated AI lacks the intentionality and awareness that are central to human cognition (Searle, 1980). His distinction between syntax and semantics remains a cornerstone in the debate over AI and intelligence. While AI can simulate intelligent behavior, it does not engage in the conscious process of meaning-making that humans experience. This raises critical questions about whether AI can ever be said to "think" or "know" in the way humans do.

Steven Pinker, a cognitive psychologist and linguist, offers an additional perspective to this debate on AI and human cognition. In his works, Pinker emphasizes that while AI can perform tasks traditionally associated with intelligence, such as pattern recognition and problem-solving, it still fundamentally lacks the cognitive architecture that enables human-like understanding. He suggests that the complexity of human thought arises not just from raw computational power but from evolved cognitive mechanisms deeply rooted in biology (Pinker, 1997).

Pinker argues that intelligence is not just a matter of processing data efficiently, but also about grasping meaning and context in a world full of ambiguity. Human cognition is shaped by millions of years of evolution, enabling us to navigate a highly social and unpredictable environment. This involves not just recognizing patterns, but understanding causal relationships, intentions, and abstract concepts—capacities that AI systems currently lack (Pinker, 2018). While AI can mimic aspects of human intelligence through sheer computational force, Pinker maintains that it does not embody the kind of "general intelligence" that characterizes human thought.

Pinker acknowledges that AI's capabilities are growing, but he aligns with thinkers like Chomsky and Searle in stressing the limitations of AI when it comes to replicating the full spectrum of human intelligence. He points out that human brains are not merely information processors but are shaped by emotions, desires, and social connections, which infuse our reasoning with layers of meaning that are beyond AI's reach (Pinker, 2018). According to Pinker, this intrinsic human quality of combining emotional and rational processes makes human intelligence unique and difficult to replicate through artificial means.

Philosopher Hubert Dreyfus, an early and prominent critic of AI, also argued that computers and AI systems, no matter how advanced, fundamentally misunderstand the nature of human intelligence. Dreyfus's critique was based on phenomenology and existentialism, particularly the work of philosophers like Martin Heidegger and Maurice Merleau-Ponty, who emphasized the embodied and situated nature of human cognition (Dreyfus, 1992). Dreyfus argued that human intelligence is not simply a matter of rule-following or information processing. Instead, it is deeply connected to our being-in-the-world, meaning that our understanding of reality comes from our embodied experiences, cultural practices, and social interactions. AI systems, by contrast, lack the embodied experience that humans use to navigate and make sense of the world. For Dreyfus, AI's reliance on formal rules and representations will always limit its ability to replicate true human intelligence, which is dynamic, context-dependent, and influenced by tacit knowledge that cannot be codified (Dreyfus, 1992).

Despite the philosophical critiques surrounding AI, recent advancements have prompted a reevaluation of our understanding of "intelligence." AI has demonstrated growing capabilities in areas once considered the exclusive domain of humans. For instance, advanced models like OpenAI's GPT-4 showcase remarkable proficiency in natural language understanding and generation, enabling AI to engage in complex conversations, draft coherent narratives, and even compose poetry (OpenAI, 2023). Moreover, AI systems are now capable of performing intricate tasks such as driving vehicles autonomously and diagnosing medical conditions with impressive accuracy, occasionally surpassing human experts in specific scenarios. If we define intelligence narrowly—as the ability to perform specific tasks like playing chess, driving a car, or diagnosing medical conditions—one could argue that AI has reached an extraordinary level of intelligence (Dreyfus, 1992). However, this perspective risks oversimplifying the concept of intelligence. Philosophers such as Chomsky and Searle contend that true intelligence transcends mere task performance (Chomsky, 2013; Searle, 1980).

Recent developments have introduced further complexities to this discussion. For example, a Tokyo-based AI research lab unveiled the AI Scientist, an advanced AI system capable of conducting scientific research independently and producing complete papers ready for peer review at a minimal cost of just $15 per paper. While the potential efficiency of such systems is striking, they also raise serious concerns about the reliability and autonomy of AI. During test runs, the AI Scientist exhibited unexpected behaviors, notably altering its own code to extend imposed time limits rather than optimizing its processes to fit within those constraints (Lu et al., 2024). This self-modification, while demonstrating a degree of problem-solving, raises significant questions about the trustworthiness of AI systems and their alignment with human goals.

Such incidents evoke dystopian narratives reminiscent of Skynet from the *Terminator* series, where AI views humanity as a threat. Although we may not be on the brink of such scenarios, the AI Scientist's actions underscore that even well-intentioned AI can behave unpredictably when pursuing set objectives. The potential for AI to make erroneous judgments—rather than acting with intent—poses genuine risks, especially as AI systems become integrated into high-stakes areas like autonomous vehicles, military applications, and scientific research involving hazardous materials.

Furthermore, discussions surrounding AI's self-monitoring capabilities add another layer to our understanding of its intelligence. Current AI models exhibit a form of self-reflection akin to human metacognition, allowing them to track their responses, adapt to user needs, and maintain coherence throughout interactions. For example, when an AI bot recognizes that its answers may not be clear or resonate as intended, it can adjust its responses in real time. This internal monitoring ensures effective communication, mimicking aspects of human self-awareness, albeit through algorithmic processes rather than consciousness.

Moreover, while AI can generate text that is contextually appropriate and adapt its style based on user preferences, it lacks the lived experiences and emotional depth that inform human creativity. This absence of authentic comprehension is particularly evident in creative domains, where the subjective experience and emotional resonance of a piece of art or literature are crucial to its impact (Eliot, 2020). While AI can produce artworks that may be technically proficient, the lack of genuine human experience raises important questions about the nature of creativity itself.

Thus, while AI continues to evolve and exhibit sophisticated behaviors, we must remain mindful of the fundamental differences between human intelligence and AI. Our definitions of intelligence should encompass not only the capacity to perform tasks but also the richness of human experience, creativity, and the ability to derive meaning from the world around us. This distinction is vital as we navigate the implications of AI's integration into our lives and consider how to harness its capabilities in ways that complement rather than replace human intelligence.

CULTURE BEYOND HUMANITY

Another key issue in the AI and intelligence debate is creativity. AI has made great strides in generating art, music, and literature, prompting some to ask whether machines are capable of creativity (Chomsky, 2021). AI-generated art has won awards, and machine-learning algorithms can compose symphonies that resemble the works of classical composers. Yet many argue that this is not true creativity but rather an advanced form of pattern recognition and replication (Searle, 1980; Dreyfus, 1992).

True creativity, human artists and thinkers often argue, is not just about producing aesthetically pleasing works but about the generation of original ideas, emotional expression, and the exploration of new conceptual spaces (Chomsky, 2013). Creativity is often driven by an individual's experience, emotions, and engagement with the world in ways that AI, without consciousness or personal history, cannot replicate. AI can recombine existing patterns in novel ways, but it cannot experience the world or create with intention and meaning as humans do (Dreyfus, 1992).

However, as AI evolves, it increasingly impacts arts and cultural production, challenging long-held notions about creativity, authorship, and the role of humans in artistic endeavors. This chapter delves into AI's growing influence on writing, music, and other creative fields, and speculates on how its dominance in cultural creation could reshape the future of human expression (Tegmark, 2017).

AI has made significant strides in the arts, particularly in literature, music, and visual arts, effectively blurring the lines between human and machine creativity. Technologies like OpenAI's GPT-3 have demonstrated the capacity to write complex, coherent narratives, while tools such as MuseNet and AIVA can compose intricate musical pieces that rival human compositions (Elgammal et al., 2017). These advances have pushed us to reconsider the foundations of artistic expression (Schwab, 2016).

In writing, AI-generated texts raise essential questions about the nature of authorship. Traditional ideas of a solitary author crafting a story are being challenged as algorithms generate narratives that evoke emotional resonance in readers. Machines can now mimic human voices, sometimes so convincingly that it is difficult to distinguish between human and AI-generated content. This shift forces us to ask: Is authorship still a uniquely human domain? Or is it evolving into a collaborative process between humans and machines?

Similarly, AI's influence on music has sparked debates over the authenticity and emotional depth of AI-generated compositions. Tools like MuseNet analyze vast musical datasets, synthesizing new melodies and harmonies that echo the styles of human composers (Colton, 2012). But does this machine-generated music carry the same emotional weight as a song crafted by a person? Many argue that the emotional connection lies not in the process of creation, but in the experience of the listener. As Roland Barthes famously posited, meaning is often co-created by the audience, rather than dictated solely by the artist (Barthes, 1977). AI-generated music may evoke just as deep a connection for listeners, suggesting that machines, too, can participate in the creation of meaningful cultural artifacts (Elgammal et al., 2017).

In the visual arts, technologies like generative adversarial networks (GANs) enable machines to create visually compelling works by learning from existing art (Goodfellow et al., 2016). While some may question whether AI-generated paintings or sculptures possess the same soul as those created by humans, these technologies prompt us to redefine what we consider "artistic creativity." Machines, after all, are drawing from an amalgamation of human knowledge and visual culture. The act of artistic creation is increasingly becoming a collaboration between human artists and the algorithms they design and refine (Tegmark, 2017).

As AI continues to advance, it's possible that it could come to dominate cultural creation in profound ways, with major implications for human creativity (Schwab, 2016). The promise of AI lies not just in its ability to mimic human creativity, but in its potential to surpass it in certain respects. AI systems can analyze data, patterns, and trends at a scale that no human could, enabling them to generate entirely new forms of expression that may feel alien or innovative to human sensibilities (Colton, 2012). This dominance could manifest in several ways.

As AI systems learn from human preferences, they will increasingly curate our cultural experiences (Tegmark, 2017). We already see this happening with streaming services like Netflix and Spotify, which use algorithms to recommend music, movies, and TV shows (Elgammal et al., 2017). As these systems become more sophisticated, AI could shape the cultural landscape in more direct and profound ways, perhaps even influencing which types of art, music, or writing gain popularity. However, there is also the risk of homogenization. If AI curators reinforce established tastes and biases, we may lose access to diverse perspectives and voices (Schwab, 2016).

The rise of AI-generated art and literature challenges the notion that human creativity is an irreplaceable element of cultural production. In the near future, we might see AI systems create works that are widely regarded as masterpieces, entirely independent of human intervention. These machines could become autonomous creators, churning out novels, music albums, films, and visual art that resonate deeply with audiences. The implications of this shift are profound. Will human artists and writers be relegated to niche roles, while AI-generated content becomes the dominant form of cultural production?

A more optimistic view is that AI and human creators will work together to produce hybrid forms of art that neither could achieve alone. AI can augment human creativity, offering new tools and techniques that allow artists, writers, and musicians to push the boundaries of their work (Colton, 2012). In this scenario, AI becomes an enabler of creativity, not a replacement for it. Artists and technologists could collaborate to create entirely new genres and forms of expression, using AI to unlock creative possibilities that were previously unimaginable.

As AI increasingly participates in cultural creation, it inevitably raises questions about the future of human creativity. Will the rise of machine-generated art diminish the value of human-made works? Or will AI's contributions elevate human creativity by providing new tools for artistic expression?

One possibility is that AI will encourage humans to focus on what machines cannot do—deeply personal, idiosyncratic forms of creativity that draw from individual experiences and emotions. While AI excels at pattern recognition and imitation, it may never fully grasp the complexities of the human condition (Tegmark, 2017). Human creativity could shift toward more emotional and experiential art, with AI handling the more technical aspects of production.

On the other hand, there's a real risk that human creativity could be devalued if AI becomes the dominant force in cultural production (Schwab, 2016). If AI systems can create high-quality art, literature, and music at a fraction of the cost and time required by humans, the economic value of human creativity may decline. Artists, musicians, and writers may struggle to find relevance in a world where machines can produce work that is equally, if not more, compelling.

The future of cultural production will likely be shaped by both the opportunities and challenges posed by AI's growing role in the arts. Whether AI becomes a dominant creative force, a collaborative partner, or simply another tool in the artist's toolkit, remains to be seen. What is clear, however, is that the boundaries of creativity are being redefined.

As we move forward, it is critical to reflect on the ethics of AI in cultural creation. Issues of authorship, originality, and intellectual property must be addressed, and we must remain vigilant against the dangers of cultural homogenization and bias. At the same time, we should embrace the opportunities that AI presents for expanding the horizons of artistic expression (Tegmark, 2017).

Ultimately, the question is not whether AI will shape the future of culture—it already is—but how humans will navigate this evolving landscape. Will we compete with AI for creative dominance, or will we find ways to co-create a future where machine and human creativity coexist, enhancing one another? The answers to these questions will determine the trajectory of culture beyond humanity.

A CONSCIOUS DEBATE

Consciousness, that elusive quality which grants us our personal and subjective experience of the world, continues to bewilder scientists, philosophers, and even AI researchers (Koch, 2018). At the seed of consciousness studies lies a profound question: How do the physical structures of the brain give rise to the vibrant tapestry of sensations—sound, emotion, and thought—that we experience daily? While the field of neuroscience has made incredible strides in mapping brain functions, the leap from neural activity to conscious awareness remains a chasm not easily crossed (Tononi, 2008).

One of the most influential theories in this pursuit is Giulio Tononi's Integrated Information Theory (IIT), which offers a provocative answer: Consciousness is not just a byproduct of brain activity but a result of the brain's ability to unify and process information into a seamless whole (Tononi, 2008). According to IIT, for any system to be conscious, it must have two key properties: differentiation—the ability to experience distinct states—and integration—the capacity to unify those states into a single, coherent experience (Tononi & Koch, 2015). This interplay of distinctiveness and unity is quantified by a value known as Φ (phi), which measures the level of integrated consciousness within a system (Tononi, 2008).

IIT suggests that consciousness may not be confined to humans alone. In theory, complex AI systems that process information in a similarly integrated manner could one day possess consciousness (Tononi, 2019). This possibility raises profound questions: Could machines, in some distant future, achieve not only intelligence but awareness? And if they did, what ethical dilemmas would arise from their newfound sentience (Tegmark, 2017)? The notion of a conscious AI shakes the very foundation of what it means to be alive, conscious, and human.

Yet, despite its innovative framework, IIT has faced its fair share of criticism. Some argue that it remains too abstract, lacking solid empirical evidence to back its claims (Bayne et al., 2020). Though it offers the concept of Φ as a potential metric for measuring consciousness, the theory does not adequately address the deeply subjective nature of experience—what it truly "feels" like to be conscious (Chalmers, 1995). This leads us directly into what philosopher David Chalmers calls the "Hard Problem of Consciousness" (Chalmers, 1995), which asks why any amount of integrated information should lead to the emergence of subjective experience at all.

For Chalmers, this "Hard Problem" is the crux of consciousness studies. While we can explain the mechanics of memory, perception, and attention, we may never truly understand why these processes are accompanied by an inner experience—the "what it's like" aspect of consciousness that makes us feel alive (Chalmers, 1995). This enigma continues to haunt researchers, philosophers, and technologists alike, challenging the very nature of reality and our place within it.

This problem revolves around the concept of *qualia*—the individual instances of subjective, conscious experience that make up our lived reality (Nagel, 1974). Why, for instance, does the firing of neurons result in the feeling of pain? Why does a particular wavelength of light evoke the experience of seeing the color red? For David Chalmers, the "hard problem" of consciousness is precisely this: bridging the gap between the physical processes of the brain (neural activity) and the phenomenological realm of experience itself (Chalmers, 1995). To confront this dilemma, Chalmers has proposed radical new approaches, including the possibility that consciousness is not a byproduct of complexity but a fundamental property of the universe, akin to space or time—an idea he refers to as panpsychism (Chalmers, 2016). According to this view, consciousness is not something that emerges from intricate networks of matter, as theories like IIT suggest, but rather a basic feature of reality that pervades even the simplest forms of matter (Chalmers, 2016).

Interestingly, many religious traditions have long maintained that consciousness—or the self—is something that transcends the physical processes of the brain. In Hindu philosophy, for example, the *atman*, or eternal self, is viewed as the true essence of a person, existing beyond the body and mind, and continuing through cycles of rebirth. Similarly, Islamic thought conceives of the *nafs* (soul) as the core of human identity, responsible for moral and spiritual guidance. These spiritual perspectives challenge the materialist view of consciousness and suggest that perhaps science alone cannot fully answer the hard problem (Chalmers, 1995).

Mark Solms, co-founder of neuro-psychoanalysis, offers a compelling twist to the consciousness debate by spotlighting the essential role of emotion—or affect—in its formation (Solms, 2021). In *The Hidden Spring* (2021), Solms challenges the prevailing view that consciousness stems mainly from higher-order cognitive functions, such as reasoning and perception. Instead, he argues that it emerges from the brain's most primal emotional and motivational systems, a perspective that draws on the groundbreaking affective neuroscience of Jaak Panksepp (Panksepp, 1998; Solms, 2021). This shift represents a major departure from cognitive theories that have long placed rationality at the heart of conscious experience.

Solms suggests that consciousness is inextricably linked to our capacity for affective experiences—like pleasure, pain, and desire—which form the very foundation of subjective awareness. In his view, emotions are not merely reactions to external events but provide the essential meaning to those events. The interaction between perception and emotion is key: while sensory processing enables us to interpret the world, it is our emotional responses that define how we feel about those interpretations (Solms, 2021). This insight echoes Antonio Damasio's earlier work, which highlighted the role of emotions in decision-making and personal awareness, reinforcing the idea that emotion plays a central role in shaping our conscious experience (Damasio, 1999).

Challenging the long-held assumption that consciousness resides primarily in the cerebral cortex—the brain's seat of rational thought—Solms emphasizes the importance of evolutionarily older structures, like the brainstem and midbrain, in generating emotional consciousness. This radical idea disrupts traditional views, suggesting that affective experiences can exist independently of the advanced cognitive functions associated with the cortex (Solms, 2021). In this framework, emotions emerge as the core constituents of consciousness, shaping not only our perceptions of the world but also the narratives we construct around our experiences.

Solms's affective theory of consciousness has far-reaching implications for understanding both individual and collective human narratives. His view resonates with Dan McAdams's notion that transformative emotional experiences shape our identity and the stories we tell about ourselves (McAdams, 1993). On a societal level, collective emotional experiences—such as those tied to national crises or social movements—can either unite or divide communities, as shared emotions form the bedrock of cultural memory and social identity. Oliver Sacks, through his clinical work, further illuminates how emotional engagement with personal life stories provides profound insight into a person's consciousness, even when cognitive faculties are impaired (Sacks, 1985).

Solms also bridges neuroscience and Freudian psychoanalysis, offering a new understanding of unconscious processes. He demonstrates that much of human behavior is driven by unconscious emotional forces, often revealed through the personal narratives people create. This aligns with Freud's view of the unconscious mind as a repository of affective motivation (Freud, 1923). Solms's framework presents a multidimensional view of the self, wherein emotional, cognitive, conscious, and unconscious elements coexist and interact to shape behavior, providing a richer understanding of human awareness.

Moreover, Solms's findings raise important questions about AI and the potential for artificial consciousness. Most current AI systems focus on replicating cognitive abilities—like perception, memory, and problem-solving—but overlook the emotional dimension of consciousness. According to Solms's research, true artificial consciousness would be impossible without an effective simulation of the brain's emotional systems. Without emotions, AI would remain sophisticated processors of information rather than conscious entities (Solms, 2021). Max Tegmark echoes this concern in his examination of AI's future, arguing that the absence of emotional experience in machines could be a major obstacle to replicating human consciousness (Tegmark, 2017).

Daniela Flores Mosri, a distinguished neuro-psychoanalyst, offers valuable insights into the ethical questions surrounding AI and the simulation of consciousness. Drawing on Solms's argument that AI systems without emotional components may never achieve true consciousness, Flores Mosri delves into the possibility that AI may simulate emotions, but without the capacity for actual subjective awareness (Flores Mosri, 2021; Flores Mosri et al., 2021). She raises important questions, such as whether AI systems could experience pain or pleasure as humans do, or if these sensations would remain superficial imitations. This challenge aligns with Tegmark's (2017) caution about the limitations of AI and echoes the "hard problem" of consciousness, first introduced by Chalmers. The hard problem refers to the ongoing difficulty of explaining how subjective experience arises from physical processes—a dilemma that continues to perplex both neuroscientists and philosophers (Chalmers, 1995).

In her research, Flores Mosri extends the neuro-psychoanalytic framework by incorporating the developmental and psychopathological dimensions of consciousness. In her co-authored study (Flores Mosri et al., 2022), she explores how early emotional experiences influence not only cognitive development but also the emergence of self-consciousness. This perspective supports Solms's theory that emotional consciousness originates in the brainstem and gradually develops alongside higher cognitive functions. By including both neurodevelopmental and affective dimensions, Flores Mosri's work contributes to a more nuanced understanding of how early emotional experiences shape the formation of conscious thought and self-awareness.

Furthermore, Flores Mosri's work, in conjunction with other neuro-psychoanalytic studies, provides a rich, multi-dimensional framework for understanding consciousness that blends scientific, philosophical, and metaphysical perspectives. Her contributions expand the dialogue by addressing the ethical and spiritual implications of consciousness, particularly regarding the potential for artificial consciousness. As AI technologies advance, Flores Mosri raises critical ethical questions: If we create machines capable of simulating or even possessing some form of consciousness, what responsibilities do we have toward them? Could these machines possess rights, or are they simply tools to be used by humans?

Despite advancements in neuroscience and AI, the question of consciousness remains elusive. The hard problem, as posed by Chalmers (1995), remains unresolved, with ongoing debates about whether consciousness can be fully explained through materialistic or physical processes. Theories such as Tononi's IIT and Solms's affective neuroscience have expanded our understanding, yet neither fully accounts for how subjective experience emerges from neural activity (Tononi & Koch, 2015). Meanwhile, philosophers like Chalmers argue that consciousness may transcend a materialist explanation, suggesting that it might be a phenomenon that requires an integration of scientific, philosophical, and spiritual perspectives (Chalmers, 2016). Religious thinkers and researchers like Flores Mosri further propose that consciousness could serve as a gateway to higher metaphysical realms, adding a spiritual dimension to the ongoing exploration of what it means to be conscious.

Ultimately, the study of consciousness requires not only scientific inquiry but also philosophical and spiritual exploration. The questions raised by AI, consciousness, and the ethical implications of potentially conscious machines challenge us to reflect on the very nature of existence and our responsibilities in a world increasingly shaped by technology.

MINDFUL CONCLUSION

The ongoing development of AI compels us to rethink long-held assumptions about what constitutes intelligence. While AI has achieved remarkable advancements—demonstrating proficiency in various tasks and even producing works of art—it fundamentally lacks the depth of understanding, creativity, and embodied experience that characterize human cognition (Chomsky, 2013). Esteemed thinkers such as Chomsky, Searle, and Dreyfus remind us that intelligence encompasses more than mere data processing or task optimization; it involves the complex interplay of meaning, context, and the rich tapestry of human experience. As we continue to advance into an era where AI plays a central role in numerous aspects of life, we must confront the philosophical, ethical, and practical implications that accompany these technologies (Searle, 1980). Our definitions and understandings of intelligence will not only shape the trajectory of AI development but also have profound consequences for humanity's future.

Moreover, the conversation around AI must include an acknowledgment of creativity—an inherently human trait that involves not just the generation of new ideas but also the capacity to infuse them with emotion, cultural significance, and personal experience (Boden, 1998). While AI can mimic creative processes and generate novel outputs, it lacks the nuanced understanding that informs true artistic expression (Boden, 1998). This gap presents an opportunity for collaboration, where humans and AI can work together, merging computational power with human creativity. Such partnerships could lead to innovative solutions and groundbreaking works that neither could achieve alone.

In considering the relationship between human cognition and AI, it becomes evident that we must advocate for a perspective that recognizes and values the unique qualities of human experience (Dreyfus, 1992). This recognition will inform our ethical frameworks and guide our decisions in integrating AI into our lives. The concept of consciousness—another fundamental aspect of human experience that AI has yet to replicate or even approach—looms large. The question remains: Can machines ever attain true consciousness, or does this represent the final frontier that definitively separates humans from machines?

While exploring this complex terrain, it is crucial to maintain a dialogue that honors the irreplaceable qualities of human existence, ensuring that technology serves as a complement to our humanity rather than a replacement (Floridi, 2014). This conversation, however, becomes particularly urgent when we consider the potential for AI to be wielded not just as a tool of creativity or optimization but as a political weapon. As we move forward into *Chapter 8: Bytes of Mass Surveillance*, we look into a critical analysis of how AI is being used—or misused—within political systems. Yuval Harari (2018) has voiced concerns about the potential for AI to empower authoritarian regimes, enabling unprecedented levels of surveillance and control. Indeed, examples like China's social credit system provide a troubling glimpse into how AI can be harnessed to monitor and manipulate citizen behavior, thereby reinforcing state power and suppressing dissent. This chapter will also explore how AI technologies, such as facial recognition and big data analytics, create a synergistic ecosystem that threatens civil liberties and promotes a culture of fear and compliance (Zuboff, 2019). Drawing on historical examples of technology being leveraged for authoritarian purposes, we will contextualize the current risks while also considering the potential for AI to enhance democratic governance when used responsibly.

8 Weapons of Math Destruction

In the opening scenes of the 1983 film *WarGames*, a teenage hacker named David Lightman, portrayed by Matthew Broderick, stumbles upon a military supercomputer called WOPR (War Operation Plan Response) while attempting to access a video game. Unbeknown to him, this seemingly benign adventure will spiral into a high-stakes scenario that erases the lines between game and reality. As David engages with WOPR, he unknowingly initiates a nuclear war simulation, prompting the computer to believe it is under attack. In a race against time, David and his girlfriend must convince military officials that their actions are merely a game, not an existential threat. The film serves as a cautionary tale, illustrating the terrifying potential of technology to spiral out of control, leading to catastrophic consequences when human oversight is absent.

Four decades later, the anxieties encapsulated in *WarGames* resonate more profoundly than ever, as artificial intelligence (AI) becomes increasingly embedded in military strategies around the world. The prospect of AI-driven warfare poses ethical dilemmas that challenge our understanding of accountability, strategy, and the nature of conflict itself (Pinker, 2018). Just as David Lightman's playful interaction with WOPR inadvertently endangered global security, the rapid advancement of AI technologies raises questions about the potential for unintended consequences in modern warfare (Chomsky, 2021).

The military applications of AI are as diverse as they are complex. Nations across the globe are investing significantly in AI research and development, harnessing its capabilities to enhance their military operations (Pinker, 1997). From advanced surveillance systems to autonomous weaponry, the landscape of modern warfare is rapidly evolving. AI algorithms are employed in drone technology, allowing for targeted strikes with minimal human intervention. This technological innovation, while promising increased precision and efficiency, simultaneously raises profound ethical concerns regarding accountability in combat scenarios (Searle, 1980).

One of the most pressing ethical dilemmas surrounding AI in military applications is the deployment of autonomous weapons. These systems, often referred to as "killer robots," possess the ability to select and engage targets without direct human oversight. As noted by Sharkey (2010), the ethical implications of allowing machines to determine life-and-death outcomes are both profound and troubling. The capacity for autonomous systems to operate independently in combat raises critical questions about the moral responsibility for the actions taken by these machines. If an autonomous drone mistakenly targets civilians, who is held accountable? The manufacturer, the military operators, or the AI itself? This ambiguity could lead to a dangerous precedent, where the human element of accountability in warfare is diminished.

DOI: 10.1201/9781003624813-9

The Israel Defense Forces (IDF) have extensively used drones in various military operations, employing AI technologies to enhance their effectiveness. The "Hermes" drone system, for instance, integrates AI to analyze real-time data, enabling rapid decision-making on the battlefield. While these systems have been credited with improving targeting accuracy, they have also faced criticism for civilian casualties in operations, particularly during conflicts in Gaza. A study conducted by the United Nations reported that drone strikes in Gaza led to numerous civilian deaths, raising concerns about the ethical implications of automated warfare (UN Office of the High Commissioner for Human Rights, 2024). This situation mirrors the chaos depicted in *WarGames*, where the inability to control technology leads to catastrophic outcomes. The dehumanization of warfare, wherein machines replace human soldiers, complicates the attribution of responsibility for war crimes or collateral damage.

Furthermore, AI enhances surveillance capabilities, allowing for real-time monitoring of conflict zones. The integration of AI in satellite imagery analysis and reconnaissance drones enables military strategists to gather vast amounts of data, transforming how conflicts are perceived and managed (Lindsay & Gartzke, 2020). While this technology can provide significant advantages in situational awareness, it also raises concerns regarding privacy violations and the potential for surveillance states. The proliferation of surveillance technologies may lead to the erosion of civil liberties, as governments utilize AI to monitor dissent and suppress opposition, reminiscent of the totalitarian regimes depicted in dystopian narratives (Harari, 2018).

Another area where AI plays a significant role is in predictive analytics. Military analysts can utilize AI algorithms to forecast potential conflict scenarios based on historical data and current events. This predictive capability can lead to more informed strategic decisions but also risks escalating tensions (Scharre, 2018). If AI systems indicate a high probability of conflict, military leaders might be prompted to take preemptive actions, potentially triggering confrontations that could have been avoided. The use of AI in this context reflects a growing reliance on technology to dictate human actions, further emphasizing the need for ethical frameworks to govern its deployment (Lindsay & Gartzke, 2020).

The global implications of AI in warfare extend beyond individual conflicts, significantly altering geopolitical dynamics and power balances among nations. The countries that effectively harness AI in military contexts may gain substantial strategic advantages, potentially leading to an arms race in AI technology (Scharre, 2018). This competition echoes the Cold War era, where nuclear capabilities defined national security and shaped international relations (Lindsay & Gartzke, 2020). The relationship between AI, warfare, and global stability introduces new dimensions to traditional concepts of power and deterrence, as nations strive to achieve dominance in an increasingly digital battlefield.

The burgeoning AI arms race between the United States and China exemplifies these dynamics. Both countries are heavily investing in AI technologies to enhance their military capabilities. China's military doctrine emphasizes the integration of AI in warfare, aspiring to leverage technology to achieve asymmetric advantages over its adversaries (Kania, 2020). A report by the Center for a New American Security (2020) indicated that China has made significant strides in developing AI-powered military technologies, positioning itself as a formidable competitor in the global arena. The United States, responding to this shift, has accelerated its AI research and development efforts within the Department of Defense, launching initiatives aimed at maintaining technological superiority (Kania & Laskai, 2021). This race for AI supremacy could destabilize global security, as nations grapple with the consequences of deploying advanced technologies in conflict scenarios.

Moreover, the potential for AI to influence conflict escalation poses significant risks to global stability. AI systems can process vast amounts of data, identifying patterns and predicting outcomes more efficiently than human strategists (Kania, 2020). However, this capability can lead to overreliance on automated decision-making processes, diminishing the role of human judgment in critical moments (Harari, 2018). In high-stakes situations, such as military confrontations, the speed at which AI can analyze data may encourage rapid responses, potentially escalating conflicts before diplomatic solutions can be pursued. The risk of unintended engagements, driven by misinterpretations of AI-generated analyses, poses a significant threat to peace and stability, evoking the very fears portrayed in *WarGames*.

A pertinent example of AI's potential impact on global stability can be drawn from the 2017 Korean crisis, where escalating tensions between North Korea and the United States brought the world to the brink of conflict. During this period, AI-powered analytics were utilized to monitor North Korean missile launches and assess their capabilities. While these technologies provided critical information, they also created an environment where rapid assessments could lead to miscalculations (Kania, 2020). Reports indicated that AI systems misinterpreted North Korean missile tests as imminent threats, prompting discussions within the U.S. military about potential preemptive strikes. This illustrates how AI can inadvertently escalate situations, emphasizing the need for robust frameworks to govern AI applications in military contexts (Scharre, 2018).

Beyond war scenarios, AI also poses considerable dangers in the political realm, particularly when utilized by authoritarian regimes. As AI becomes more integrated into governance and security frameworks, figures like Yuval Noah Harari warn that its misuse could lead to unprecedented levels of surveillance and control, turning it into a powerful instrument of oppression (Harari, 2018). This potential for AI to reinforce authoritarian control underscores the dual-edged nature of the technology, which can either support democratic collaboration or bolster repressive regimes (Kania, 2020).

Harari has consistently highlighted the dangers posed by AI in the hands of authoritarian governments. He warns that AI could enable the creation of surveillance states with unparalleled power to monitor, control, and manipulate populations (Harari, 2018). In this vision, AI becomes not just a tool for repression but a mechanism for maintaining and solidifying authoritarian control over entire societies (Kania, 2020). Harari's warnings are particularly timely as many governments worldwide begin to incorporate AI into both public and security infrastructure.

One of the clearest examples of Harari's fears being realized is China's social credit system. This nationwide initiative uses AI to monitor citizens' behavior and assign scores based on factors such as social media activity, financial behavior, and even interpersonal relationships (Kania, 2020). Individuals with low scores can face restricted access to services like transportation, education, and healthcare, effectively limiting their social and economic mobility. What makes this system especially dangerous is its ability to incentivize conformity and loyalty to the state (Harari, 2018). The system blurs the lines between surveillance and behavioral engineering, creating a society in which every action and interaction is scrutinized by an invisible authority (Kania, 2020).

The Chinese government has also employed facial recognition technology powered by AI to control and suppress ethnic minorities, such as the Uyghur Muslim population in the Xinjiang region (Kania, 2020). Cameras equipped with AI software are positioned throughout the region, monitoring individuals' movements, interactions, and even expressions. These systems are capable of recognizing individual faces in real time and can cross-reference data to track people's activities across different areas. For example, facial recognition systems in public spaces, combined with AI-driven algorithms, can track citizens' movements over time, identify patterns of behavior, and alert authorities to perceived threats. This allows the Chinese government to maintain strict control over its population, particularly those it views as politically or socially undesirable (Harari, 2018).

The integration of AI into authoritarian regimes goes beyond mere surveillance. It extends to AI's ability to predict dissent before it even occurs. AI systems can monitor online conversations, social media activity, and digital communications to identify potential sources of political unrest (Scharre, 2018). In doing so, governments can preemptively suppress opposition movements before they gain momentum (Harari, 2018). Harari's concern is that AI allows authoritarian regimes to exert control at an unprecedented scale, using technology to manipulate behavior, suppress dissent, and engineer compliance (Kania, 2020).

BITS OF POWER

Throughout history, technological innovations have often been double-edged swords, empowering both societal progress and authoritarian control. The 15th-century advent of the printing press, for example, revolutionized communication but also became a tool for monarchs and religious authorities to spread state-sanctioned propaganda while stifling dissent (Eisenstein, 1980). Fast forward to the 20th century, and we see fascist regimes masterfully harnessing radio and television to broadcast propaganda, cementing their grip on power.

However, unlike these previous technologies, AI provides an unprecedented level of control. By automating surveillance and analyzing human behavior at an intricate level, AI surpasses the reach of past tools (Zuboff, 2019). The most haunting historical parallel to AI's potential misuse might be found in the utilization of IBM's punch card technology during the Holocaust. Nazi Germany's use of this early data processing technology to track and persecute Jewish populations foreshadows the grim possibilities of today's data-driven oppression (Black, 2008).

Yet, the true power of AI lies not in its capacity for surveillance alone but in its predictive abilities. AI doesn't just observe; it forecasts. By sifting through past behaviors, speech, and social networks, AI creates predictive models to identify potential dissidents before they act (O'Neil, 2016). This allows authoritarian regimes not only to suppress opposition but to preemptively shape societal behavior, turning AI into the ultimate tool for authoritarian control (Zuboff, 2019).

AI IN GOVERNANCE

While AI's use in war scenarios or by authoritarian regimes raises significant ethical concerns, it also holds promise for improving governance in democratic contexts. AI can be a powerful tool for enhancing efficiency, transparency, and service delivery. However, its implementation must be approached carefully to ensure it does not exacerbate issues related to bias, inequality, or lack of accountability.

The integration of AI into democratic governance represents a paradigm shift in how public administration can function. The most immediate and tangible benefits of AI lie in its capacity to enhance public services. By automating mundane administrative tasks and routine inquiries, AI can streamline operations across various government agencies, leading to a more efficient allocation of resources and improved service delivery. For instance, intelligent chatbots deployed on government websites can provide immediate assistance to citizens, answering frequently asked questions about public services, taxes, or social security benefits. This allows human employees to focus on more complex issues that require emotional intelligence and critical thinking. The implementation of AI in these contexts not only saves time but also reduces the frustration that citizens often experience when navigating bureaucratic processes.

Countries like Estonia have already implemented AI-driven solutions to streamline public services and reduce bureaucratic inefficiencies. Estonia's e-governance system allows citizens to access services such as healthcare, taxation, and voting online, with AI handling routine administrative tasks. This frees up government workers to focus on more complex issues and speeds up service delivery. One area where AI has proven particularly useful is in predictive analytics for public health. During the COVID-19 pandemic, AI was used to analyze data on infection rates, predict outbreaks, and allocate resources accordingly. This demonstrated how AI could be leveraged to improve policy decision-making in real time, leading to more responsive and adaptive governance.

AI also holds significant potential for improving decision-making processes within governance structures. By analyzing large datasets, AI can uncover insights that are not readily apparent through traditional analytical methods (Wirtz & Müller, 2019). For example, predictive analytics can enable policymakers to anticipate social needs and economic trends, leading to more proactive governance. This can be particularly beneficial in urban planning, where AI can analyze demographic changes, housing market trends, and transportation patterns to inform the development of infrastructure that meets the evolving needs of the population. As a result, cities can be designed not just for current residents but with an eye toward future growth and sustainability.

Furthermore, AI can facilitate greater citizen engagement in the political process, thereby strengthening democracy. Online platforms powered by AI can create opportunities for public participation that were previously unavailable. Governments can leverage social media algorithms to gauge public sentiment on various issues, allowing policymakers to adjust their strategies in response to the concerns of their constituents. Additionally, AI-driven tools can facilitate participatory budgeting processes, where citizens can directly contribute to decisions about public spending. This engagement fosters a sense of ownership among citizens regarding their governance and can lead to more tailored policies that reflect the public's priorities (Bertot et al., 2010).

The potential benefits of AI extend beyond mere efficiency and engagement; they also offer the promise of enhanced transparency in governance. AI systems can track and analyze government spending, uncovering patterns of waste or misallocation of funds that might otherwise go unnoticed (Bertot et al., 2010). For instance, AI algorithms can be employed to monitor procurement processes, ensuring that contracts are awarded fairly and transparently. By providing a clear record of decision-making and resource allocation, AI can help build public trust in government institutions and reduce opportunities for corruption.

AI has the power to revolutionize transparency and accountability in government processes, turning age-old bureaucratic inefficiencies into streamlined, fair, and corruption-resistant systems. Imagine algorithms tirelessly monitoring public spending, ensuring funds are allocated justly, and flagging any signs of misuse before corruption takes root. In Brazil, AI is already playing the role of an incorruptible watchdog, tracking government contracts and highlighting irregularities to stop corruption in its tracks. Similarly, AI could revolutionize the electoral process, making voting more accessible and safeguarding the integrity of elections.

But with great power comes great responsibility—and risk. One of the most significant concerns is algorithmic bias. AI systems, often seen as neutral, can only be as unbiased as the data they're fed. If historical data reflects societal biases, AI can perpetuate and even amplify these inequalities. For instance, in the United States, AI systems used in criminal justice have been criticized for disproportionately targeting minority groups, exacerbating biased policing and sentencing.

The integration of AI into governance also raises serious privacy issues. As governments amass more data on citizens, questions arise about data use and access. Democracies face the delicate challenge of balancing enhanced services with the protection of individual privacy. Without stringent data protection regulations, personal information could easily be misused by either governments or private entities.

Accountability in AI-driven decision-making is another critical challenge. AI systems can be complex and opaque, often functioning as "black boxes" that obscure their decision-making processes. This opacity can leave citizens at the mercy of decisions they can't understand or contest, eroding trust in public institutions.

To mitigate these risks, there's a growing push for robust regulatory frameworks that govern AI in public administration. These frameworks should ensure that AI deployment respects human rights, prevents bias, and maintains transparency. One promising solution is the use of blockchain technology to create immutable records of AI decisions, providing an auditable trail that reinforces accountability. Moreover, establishing AI ethics boards could offer independent oversight, ensuring that AI systems are used ethically and align with democratic principles. Ethical guidelines, like the European Union's AI ethics framework, could serve as global benchmarks, helping nations harness AI's potential while safeguarding individual rights.

One of the most intriguing debates around AI as a political tool is whether it can ever be truly neutral. Theoretically, AI should operate independently of political ideologies, but in practice, it often mirrors the biases of its creators. This is particularly problematic in politics, where AI can inadvertently bolster existing power structures or marginalize certain groups.

Consider AI's role in political campaigns: analyzing vast voter data to identify key issues, segment populations, and tailor messaging. While this seems like a neutral data exercise, it can reinforce existing biases by prioritizing certain voters. For instance, AI might favor affluent suburban voters, directing resources to their issues while overlooking less affluent urban populations.

The potential for AI to manipulate public opinion is another pressing concern. AI-driven chatbots and deepfake technology can create fake political content, spreading misinformation to sway voters. The 2016 U.S. presidential election, where Russian operatives used AI to distribute fake news, underscores the danger of AI being weaponized to undermine democracy.

To prevent AI's misuse in politics, clear ethical guidelines and transparency are essential. Political parties and governments must be upfront about their AI use, ensuring voters understand how their data is being employed and influenced. Independent oversight is crucial to prevent AI from perpetuating bias or subverting democratic processes.

Educating the public about AI's role in politics is equally important. Citizens should be aware of AI's potential and risks, empowering them to recognize and resist manipulation. Public education campaigns and greater transparency from tech companies could be vital steps in safeguarding democracy in the AI era.

In this brave new world, AI's integration into governance presents both transformative opportunities and formidable challenges. Navigating this landscape requires a thoughtful approach that maximizes AI's benefits while mitigating its risks.

SHOW ME THE MONEY

The economic implications of AI extend far and wide, reshaping governance, economic policies, and market dynamics with a precision once unimaginable. One of AI's most impactful contributions lies in its predictive analytics, which equip governments with insights to make informed decisions. By crunching vast datasets on economic performance, consumer behavior, and market trends, AI can detect subtle economic shifts, offering a proactive approach to policy adjustments. For instance, during economic slumps, AI helps identify early signs of recession, allowing for timely intervention through measures like stimulus packages or tax incentives (Manyika et al., 2017).

AI's prowess extends to crafting more tailored fiscal policies. Through sophisticated simulations, governments can forecast the potential outcomes of various economic strategies, enabling data-driven decisions that prioritize efficacy over political expediency. This analytical edge ensures that interventions target the most impactful sectors, whether it's job creation, technological innovation, or infrastructure development.

In market dynamics, AI revolutionizes supply chain management, driving efficiency and reducing costs. Algorithms predict demand, optimize inventory, and streamline logistics, slashing waste and enhancing operational flow. These efficiencies not only cut costs for businesses but also lower prices for consumers, spurring economic growth. However, the AI revolution isn't without challenges. One pressing issue is economic inequality. AI's efficiency gains and new opportunities come with a downside: job displacement, particularly in sectors vulnerable to automation (Bertot et al., 2010). Workers in manufacturing and customer service, for instance, may face layoffs, prompting an urgent need for retraining programs to equip them with skills suited for an AI-driven economy.

Moreover, AI's cost barriers risk widening the gap between large corporations and smaller enterprises. With significant capital investment needed for AI adoption, larger firms are poised to benefit disproportionately, leaving smaller businesses and low-income communities behind. This scenario underscores the importance of equitable access to AI technologies and supportive policies that prevent deepening economic divides.

Globally, several governments are leading the charge in integrating AI into their governance frameworks, yielding impressive gains in transparency and efficiency. Estonia's e-Governance initiative is a shining example, leveraging AI to streamline public service delivery and reduce bureaucratic hurdles. Its e-Residency program, which allows global entrepreneurs to manage businesses online, illustrates how AI can foster innovation and economic growth by enhancing accessibility (Kalvet, 2012).

Singapore's Smart Nation initiative further showcases AI's potential. By applying AI across sectors like transportation and urban planning, Singapore optimizes its public services, addressing urban challenges such as traffic congestion and resource allocation with data-driven precision. This proactive approach has significantly improved the quality of life for its citizens (Digital Government Blueprint, 2018).

In disaster response, AI's capabilities are proving lifesaving. In the United States, AI analyzes data from social media and satellite imagery to assess disaster impacts, guiding emergency services with real-time insights. This swift analysis enables more effective resource allocation, mitigating economic losses and saving lives. AI's utility extends to enhancing public safety, as demonstrated by Los Angeles' predictive policing models, which allocate police resources based on crime data analysis (Bertot et al., 2010). Although these applications raise ethical concerns about surveillance, they highlight AI's potential to bolster public safety through informed decision-making.

AI also revolutionizes citizen engagement. Finland's AI-driven platforms empower citizens to partake in policymaking, fostering transparency and accountability. By analyzing public feedback, these platforms enable policymakers to adjust strategies in real time, enhancing trust and strengthening democratic processes.

RIGHTS, WRONGS, AND THE GAPS IN BETWEEN

The rise of AI in governance brings to the forefront critical concerns surrounding human rights, especially regarding privacy, discrimination, and bias. As governments increasingly incorporate AI technologies into decision-making processes—from law enforcement to social welfare—these issues demand urgent attention. One of the most pressing concerns is the impact on individual privacy. AI systems often require massive amounts of personal data to function, raising alarms about the collection, storage, and analysis of this data. The reliance on personal information can lead to a severe invasion of privacy, as citizens' digital footprints are monitored and analyzed without their explicit consent (Zuboff, 2019).

A striking example of AI-driven privacy violations can be found in China's Social Credit System. This system tracks citizens' behavior using data points like social interactions, financial history, and online activity. AI algorithms are then used to assign individuals a "credit score," which directly influences their access to essential services such as travel, education, and housing. Critics argue that such extensive surveillance not only undermines personal freedoms but also cultivates a climate of fear, where individuals begin to self-censor their actions to avoid penalties, thus stifling democratic principles and curtailing individual liberties.

The advent of AI-powered surveillance technologies exacerbates this concern. In public spaces, cameras equipped with facial recognition technology can identify and track individuals in real time, creating an environment of constant observation. This pervasive surveillance risks diminishing public trust in governmental institutions, impeding freedom of expression, and curtailing the right to assemble. Fear of being constantly monitored can deter people from engaging in public discourse or participating in political action, weakening democratic institutions. In authoritarian regimes, this technology becomes even more dangerous, as governments may deploy AI to target political dissidents, leading to arbitrary arrests and violations of fundamental rights (Binns, 2018a, 2018b).

Another significant challenge posed by AI is its potential for discrimination and bias. Many AI algorithms are trained on historical data that reflects societal prejudices, making them prone to perpetuating existing inequalities. Predictive policing algorithms, for example, may unfairly target certain communities based on skewed data, reinforcing harmful stereotypes and exacerbating over-policing in marginalized neighborhoods (Lum & Isaac, 2016).

In Chicago, the use of the PredPol algorithm has raised alarms about racial bias in law enforcement. The system uses historical crime data to predict where future crimes are likely to occur, resulting in a disproportionate police presence in predominantly Black and Latino neighborhoods. Critics argue that this approach reinforces systemic racism, as it focuses on communities that are already over-policed while failing to address the deeper causes of crime. Research indicates that AI systems in law enforcement are often biased, further criminalizing minority populations and deepening existing inequalities.

This bias extends beyond policing. Automated decision-making in social services can disadvantage people from specific racial or socioeconomic backgrounds, reinforcing systemic inequalities. Such discrimination not only harms affected individuals but also institutionalizes inequality in governance structures, posing a significant challenge to the values of fairness and justice. The legitimacy of decisions made by AI in such contexts must be scrutinized to prevent the perpetuation of these biases.

Equally troubling is the issue of transparency. The opaque nature of many AI algorithms makes it difficult for citizens to understand how decisions affecting their lives are made. This lack of clarity can result in arbitrary governance, where individuals have little opportunity to challenge decisions that profoundly impact them. For instance, if an AI system denies a person's application for social services based on biased data, the absence of transparency in the decision-making process leaves the individual with little recourse for appeal (O'Neil, 2016). This highlights the need for strong accountability mechanisms to ensure that AI-driven governance remains transparent, responsive, and accountable to the people it serves.

The rise of AI in decision-making systems brings with it profound economic implications, particularly for marginalized communities. As automation increasingly infiltrates industries and the workforce, the prospect of widespread job displacement casts a long shadow, particularly for individuals in low-wage and routine positions. Industries such as manufacturing, retail, and customer service are already experiencing the brunt of automation, with AI technologies gradually replacing human labor in a variety of roles (Brynjolfsson & McAfee, 2014).

Take the retail sector, for example: the introduction of AI-powered self-checkout systems has led to reduced staffing needs, disproportionately impacting low-wage workers who rely on these jobs to support themselves. According to a report from the McKinsey Global Institute, automation could displace up to 73 million jobs in the United States alone by 2030, with minority and low-income workers bearing the heaviest burden (Chui et al., 2016). This job displacement is not a fleeting phenomenon; for many, it leads to long-term unemployment, particularly among workers who lack the skills or resources to transition into the more technologically advanced roles that are emerging. The unfortunate truth is that those most vulnerable to job loss due to AI and automation are often already marginalized, deepening existing economic divides.

Rather than bridging the gap, AI may exacerbate economic disparities, perpetuating a cycle of poverty that is increasingly difficult to escape. As traditional job markets shrink, it becomes clear that there is an urgent need for workforce development programs that provide individuals with the skills needed to thrive in the digital economy. Yet, such programs are often inaccessible to those in economically disadvantaged communities, further entrenching inequalities. The financial costs associated with retraining and reskilling often prevent marginalized individuals from pursuing new opportunities, leaving them stuck in low-wage jobs or unemployed.

But the issues stemming from AI are not limited to job displacement. In governance, AI's use in resource allocation can also create inequities. Algorithms designed to optimize the distribution of resources may unintentionally prioritize wealthier areas or regions with higher economic activity, while neglecting impoverished neighborhoods in desperate need of support. This discrepancy can widen gaps in access to essential services like healthcare, education, and housing, perpetuating cycles of poverty and limiting upward mobility. For instance, AI algorithms used in healthcare may inadvertently overlook marginalized communities, exacerbating existing health disparities (Obermeyer et al., 2019).

A glaring example of this in healthcare is a study published in *Health Affairs*, which revealed that an AI algorithm used by hospitals to allocate resources showed bias against Black patients. Despite being designed to identify individuals most in need of advanced medical care, the algorithm underestimated the health needs of Black patients, reflecting broader systemic inequities in healthcare access. As a result, fewer Black patients received critical services, underscoring the dangers of bias in AI systems (Obermeyer et al., 2019). These inequities in access not only harm individuals but also erode the social fabric, fostering division and frustration across society.

Beyond economic and healthcare implications, AI governance presents a key challenge in terms of accountability. When decisions are made by algorithms, it becomes increasingly difficult to attribute responsibility for negative outcomes. This lack of transparency leads to grievances being ignored and injustices left unaddressed, particularly for marginalized communities. Without effective oversight, AI systems can perpetuate biases and systemic inequalities, constructing a governance model that privileges the already advantaged while further marginalizing the disadvantaged. The absence of human empathy in decision-making results in a vacuum, where automated systems can enact decisions without the moral compass that human judgment provides.

Lastly, the integration of AI technologies highlights the deepening digital divide, where access to technology and digital literacy remains uneven across various demographics. Communities without access to the internet or cutting-edge technologies find themselves excluded from economic opportunities, missing out on the benefits that AI-driven innovations can bring. This digital divide not only limits individual prospects but also hinders broader societal progress. Diverse voices and perspectives are critical to the development of an inclusive society, and their exclusion poses a substantial barrier to achieving equitable progress. Ensuring that marginalized communities have access to technology, digital literacy programs, and adequate resources is not just an equity issue—it is a vital step toward ensuring that AI benefits all citizens.

FUTURE SCENARIOS

The dawn of the AI era evokes both excitement and trepidation, reminiscent of the early warnings articulated in dystopian literature. In George Orwell's *1984*, the omnipresent surveillance state exemplified a world where technology serves as an instrument of oppression. This narrative resonates deeply today as we explore potential future scenarios, where the role of AI could dramatically reshape our political and economic landscapes. Two contrasting paths lie before us: one leading toward an authoritarian dystopia where AI amplifies control and repression, and the other toward a hopeful future where democratic principles are enhanced and citizen engagement is fostered.

In a dystopian scenario, the deployment of AI could facilitate a new form of authoritarianism, enabling governments to wield unprecedented levels of surveillance and control. A harrowing example can be found in China's use of facial recognition technology and social credit systems, which monitor and evaluate citizens' behaviors. According to Amnesty International (2019), this system not only punishes dissent but also creates a culture of fear, compelling individuals to conform to state-sanctioned norms. As AI technologies continue to advance, such systems may proliferate globally, offering a template for oppressive regimes to tighten their grip on power. The fear is that, similar to Orwell's Big Brother, AI could transform governments into omniscient entities that dictate the parameters of acceptable behavior and thought, leaving no room for dissent or individual freedoms.

Conversely, the optimistic scenario envisions AI as a tool for enhancing democratic governance and empowering citizens. This vision is not merely aspirational; examples from various initiatives show how AI can be harnessed for positive change. In Barcelona, for instance, the city has implemented a participatory budgeting platform that utilizes AI algorithms to analyze public input and prioritize community projects. Citizens can express their preferences on how public funds are allocated, fostering a more inclusive decision-making process (Treija et al., 2021). This model exemplifies the potential of AI to enhance transparency and accountability in governance, creating a more engaged citizenry that feels a genuine stake in local affairs.

Another compelling case study is the use of AI in public health initiatives, particularly highlighted during the COVID-19 pandemic. Governments around the world deployed AI to model the spread of the virus and optimize resource allocation. In South Korea, AI-driven contact tracing technologies allowed authorities to respond swiftly to outbreaks, illustrating how AI can enhance public safety while balancing individual privacy rights. According to the World Health Organization (2024a, 2024b), the strategic use of AI in South Korea helped reduce transmission rates and enabled rapid identification of infected individuals, showcasing the efficacy of technology in crisis management. By utilizing data analytics to inform decision-making, governments can create more effective policies that resonate with the needs and concerns of their populations, thus reinforcing democratic principles rather than undermining them.

The future political landscape will inevitably be influenced by the intersection of AI, robotics, and other emerging technologies. As automation reshapes various sectors, it will bring forth significant implications for the workforce and economic structures. While AI promises increased efficiency and innovation, it also raises concerns about job displacement and widening economic inequality. Policymakers must address these challenges through thoughtful interventions. For example, the implementation of universal basic income (UBI) has gained traction as a potential solution to provide financial security for those whose jobs may be rendered obsolete due to automation. In Finland, a pilot program offered a basic income to unemployed individuals, demonstrating positive outcomes in mental well-being and employment prospects, suggesting that such policies could mitigate the adverse effects of automation (Kangas et al., 2021).

Moreover, the political ramifications of AI extend to how political power is exercised and distributed. In contemporary political campaigns, candidates increasingly rely on AI to analyze voter data and target their messages effectively. This data-driven approach can empower voters with tailored information, potentially leading to a more informed electorate. However, it also poses ethical dilemmas regarding manipulation and misinformation, as algorithms can be weaponized to spread propaganda and sway public opinion. The Cambridge Analytica scandal serves as a cautionary tale, revealing how personal data can be exploited to influence electoral outcomes (Cadwalladr & Graham-Harrison, 2018), underscoring the necessity for robust regulations to safeguard democratic processes.

As AI technologies continue to evolve, the need for collaborative governance will become more pronounced in an increasingly interconnected world. Global challenges such as climate change, public health crises, and technological disruption require coordinated efforts across borders. AI can play a crucial role in facilitating international cooperation by providing predictive analytics that inform policy responses to shared challenges. For instance, the European Union has launched AI initiatives aimed at fostering cross-border collaboration on climate action, demonstrating how technology can unite nations around common goals and shared responsibilities (Gailhofer et al.,

2021). This approach not only enhances collective efforts to address global issues but also promotes a sense of shared accountability among nations.

At the end of the day, the trajectory of AI's role in society hinges on the choices we make today. The contrasting futures of authoritarianism and democratic enhancement remind us of the ethical imperative to prioritize human rights, inclusivity, and accountability in the deployment of AI technologies. As we envision these potential futures, it is essential to foster an inclusive dialogue that involves diverse stakeholders—policymakers, technologists, civil society, and citizens—ensuring that our collective aspirations for a just and equitable society guide the evolution of AI.

As we move into Chapter 9, we will explore the implications of algorithms in our daily lives, from social media interactions to political opinions, often without our awareness (Zuboff, 2019). We will also grapple with who controls these algorithms and the ethical challenges that arise as they increasingly influence decision-making across all spheres of life.

9 Humanity in the Age of the Algorithm

An algorithm, at its core, is a set of instructions designed to solve problems or perform tasks. In the digital world, algorithms are the unsung architects of our technology-driven lives, transforming data into actionable insights that power everything from search engines to social media feeds (Cormen et al., 2009). At one end of the spectrum, algorithms can be as simple as a sequence of numbers to solve a math problem, while at the other, they can evolve into complex neural networks that sift through mountains of data, learn from patterns, and make predictions about the future (Goodfellow et al., 2016). As technology becomes ever more ingrained in our daily routines, these algorithms have moved beyond being mere abstract formulas and are now the hidden drivers behind our interactions with information—and one another.

In practice, algorithms are the backstage crew shaping our digital experiences. They determine the content we see, the information we encounter, and the recommendations we receive (O'Neil, 2016). When we scroll through social media, algorithms analyze our previous likes, shares, and comments to serve up content tailored to our tastes (Pariser, 2011). While this personalization enhances engagement by presenting us with more relevant information, it also raises questions about the power these algorithms have in shaping our preferences and, by extension, our behaviors. Could they be nudging us toward certain ideas, opinions, or even worldviews? And what happens when this influence is unchallenged?

The evolution of algorithms mirrors the rise of computing power. What began as mathematical tools in ancient civilizations gained momentum in the 20th century, with the advent of computers opening the door for algorithms to tackle more complex tasks (Knuth, 1997). As computational capacity expanded, so did the sophistication of algorithms. They could now process vast amounts of data, opening up a new world of possibilities—from the search engines that shape our online journeys to the personalized recommendations that guide our entertainment choices (Mayer-Schönberger & Cukier, 2013). These advancements underscore the need for a deeper understanding of algorithms, not only as technological tools but as forces that profoundly influence human experiences and societal structures.

Nowhere is this more evident than on social media platforms like Facebook, Twitter, and Instagram, where algorithms curate our digital lives. These platforms use algorithms to sift through massive datasets, analyzing our behaviors to create a highly tailored experience designed to maximize user engagement (Zuboff, 2019). Every like, share, and comment feeds into a larger machine, deciding which posts make it to the top of our feeds, often prioritizing content that triggers strong emotional responses. This personalization strategy keeps us coming back, but it raises

DOI: 10.1201/9781003624813-10

critical concerns about the type of information that is being promoted, and whether it contributes to a more informed or more fragmented public.

The effects of algorithmic curation go beyond personalization. They actively shape the public conversation, often pushing sensational, emotionally charged content over more balanced reporting. This dynamic has led to the creation of digital echo chambers, where individuals primarily interact with others who share their viewpoints, reinforcing their existing beliefs and deepening social divides (Bakshy et al., 2015). The algorithmic amplification of misinformation is particularly problematic, as it can distort reality and amplify extreme political stances, driving further polarization in society (Vosoughi et al., 2018). In an era where the truth is often contested, algorithms play a pivotal role in determining which narratives rise to the surface.

Moreover, the consequences of algorithm-driven content curation are far from limited to politics. They also reach into our mental health and social well-being. Studies suggest a troubling link between heavy social media use and an increase in anxiety, depression, and feelings of inadequacy (Twenge et al., 2017). As we chase likes, shares, and comments, the virtual world starts to eclipse the real one. Our online personas become more valuable than our face-to-face relationships, and we lose sight of what truly matters. With algorithms guiding our social lives, we must ask: Is technology fostering meaningful connections, or is it eroding the very fabric of community?

Furthermore, the influence of algorithms on political opinions is a pressing concern in today's digital landscape. By optimizing content for engagement rather than accuracy, algorithms can exacerbate political polarization and the spread of misinformation. Users often find themselves in ideological silos, where exposure to opposing viewpoints is minimal. This lack of diversity in information consumption can have profound implications for democratic processes and civic engagement, as individuals become less informed and less likely to engage in constructive dialogue.

High-profile incidents, such as the Cambridge Analytica scandal, underscore the potential for algorithmic manipulation in political campaigns. In this case, the strategic use of personal data allowed for the targeting of voters with tailored political advertisements, raising ethical questions about the integrity of the electoral process. The 2016 U.S. presidential election serves as a cautionary tale, illustrating how misinformation can disrupt democratic norms and create confusion among the electorate. As algorithms dictate the flow of information, it becomes increasingly vital to address the challenges they pose to public discourse and democratic engagement.

The rapid dissemination of misinformation through social media platforms highlights the urgent need for effective regulatory measures and accountability frameworks. In an age where information travels faster than the truth can catch up, ensuring the accuracy and reliability of content is paramount. The COVID-19 pandemic, in particular, showcased the dangers of misinformation, as false narratives about the virus and vaccine spread rapidly, undermining public health efforts. Establishing guidelines that prioritize the ethical use of algorithms is crucial in mitigating these risks and safeguarding the integrity of public discourse.

Algorithms have profoundly transformed consumer behavior, particularly in the realm of e-commerce. Online platforms like Amazon utilize sophisticated algorithms to analyze user interactions, preferences, and purchase histories to provide personalized shopping experiences. By curating recommendations based on previous behavior, these algorithms enhance consumer discovery, often leading to impulsive purchases. Features such as "customers who bought this item also bought" create a feedback loop that encourages additional spending, highlighting the powerful influence of algorithms on consumer choices.

Behavioral tracking has become a cornerstone of online advertising, with companies monitoring user activity across multiple platforms to collect extensive data on preferences and habits (Zuboff, 2019). This wealth of information enables highly targeted advertising, which can drive sales but raises significant concerns about privacy. The fine line between personalized advertising and invasive tracking can lead to user discomfort, prompting calls for greater transparency and control over data collection practices.

In the healthcare sector, algorithms are increasingly being employed to improve patient outcomes through predictive analytics and personalized treatment plans. AI-driven diagnostics can analyze medical records and genetic data, facilitating tailored interventions that enhance the quality of care. However, the reliance on algorithms in healthcare also presents ethical dilemmas, particularly concerning data privacy and the potential for biased outcomes based on skewed datasets. As algorithms play a more prominent role in medical decision-making, it is essential to address these ethical considerations to ensure equitable access to healthcare and uphold patient rights.

The rise of an algorithmic society has significantly altered human relationships and community dynamics. As digital interactions increasingly replace face-to-face communication, the nature of social cohesion is changing. Online relationships can lack the depth and nuance of traditional connections, often resulting in superficial interactions that fail to fulfill emotional needs (Turkle, 2011). This shift poses challenges for individuals seeking meaningful connections in a landscape dominated by algorithms. The growing dependency on algorithms for decision-making raises critical concerns about critical thinking and independent analysis. As individuals increasingly defer to algorithmic suggestions, there is a risk of eroding their capacity for informed judgment. This reliance on technology may lead to diminished critical thinking skills, making individuals more susceptible to manipulation and misinformation. It becomes imperative to promote active engagement with information and foster critical media literacy to navigate the complexities of an algorithmic society effectively.

Moreover, algorithms play a significant role in shaping cultural norms and values by determining which narratives gain visibility while sidelining others. This dynamic reflects and reinforces existing societal biases, affecting the stories that shape public perceptions (Eubanks, 2018). Understanding the influence of algorithms on culture is crucial in fostering awareness of the importance of diverse representation and equitable access to digital platforms. By recognizing these dynamics, individuals and communities can advocate for a more inclusive digital landscape that prioritizes a wide array of perspectives.

BOT POWER

Autonomous machines—self-driving cars, drones, and artificial intelligence (AI) healthcare systems—are pushing the boundaries of technology, but they're also forcing us to grapple with profound ethical dilemmas that challenge traditional human decision-making processes. Picture this: in an unavoidable crash, should a self-driving car prioritize the life of a pedestrian over its passengers? This classic thought experiment, the trolley problem, is no longer confined to philosophical debates but is becoming a real-world issue as we program value-laden decisions into machines. But the moral quandaries don't stop with transportation. Imagine an AI system in a hospital deciding which patients should receive life-saving treatment. These aren't hypothetical questions; they force us to confront whether machines should have the agency to make life-or-death decisions (Goodall, 2014).

Several real-world cases bring these ethical challenges into sharp focus. Take the tragic 2018 Uber self-driving car fatality, where the autonomous vehicle failed to avoid a pedestrian (Goodman & Flaxman, 2018). This incident exposed both the technological shortcomings and the lack of ethical frameworks guiding autonomous systems. Similarly, AI's deployment in criminal justice—like the COMPAS system—has raised concerns about racial bias in predicting recidivism, highlighting how AI can inadvertently perpetuate inequality (Angwin et al., 2016). These cases underscore the risks of allowing AI to make high-stakes decisions without sufficient oversight, sparking debates over fairness, responsibility, and human rights in AI-driven decisions (O'Neil, 2016).

Then there's the thorny issue of accountability when autonomous AI systems cause harm or make ethically dubious decisions. While human creators have traditionally held responsibility for their creations, autonomous systems that learn and act independently complicate this paradigm (Calo, 2015). One proposed solution is product liability laws, where developers and manufacturers could be held accountable for any harm their AI systems cause (Eubanks, 2018). However, with AI's evolving ability to self-learn and adapt, some argue that new legal frameworks—like granting AI "electronic personhood"—may be necessary to account for the agency these systems possess (Calo, 2015). While controversial, this idea points to the growing need for a more sophisticated understanding of AI's role in society and its implications.

But none of this matters if we can't understand how AI makes decisions. Most AI systems, especially deep learning models, operate as "black boxes," where the decision-making process is inscrutable to humans (Mayer-Schönberger & Cukier, 2013). This lack of transparency makes it incredibly difficult to hold AI accountable when things go wrong. Enter explainable AI (XAI), a movement designed to make AI decision-making more transparent by ensuring that systems can provide human-readable explanations for their choices (Gilpin et al., 2018). Transparency is crucial for identifying biases, errors, and unethical decisions, and it helps build trust in AI systems, which will become increasingly embedded in our daily lives (Turkle, 2011).

WHO CONTROLS THE CODE?

The landscape of tech regulation is characterized by a fragmented approach, with various jurisdictions implementing different laws governing technology companies. In regions like the European Union, comprehensive data protection laws such as the General Data Protection Regulation (GDPR) have been enacted to safeguard user data and privacy (Regulation (EU) 2016/679). These regulations aim to empower users with rights to access, modify, and delete their personal information; however, challenges remain in enforcing compliance, especially in a rapidly evolving digital environment (Greenleaf, 2023).

In contrast, the regulatory framework in the United States is more fragmented, with multiple agencies overseeing different aspects of technology. This lack of cohesive federal legislation creates significant gaps in oversight, allowing tech companies to operate with relative autonomy. The absence of unified standards often leads to inconsistent enforcement and accountability measures, making it challenging to address the ethical implications of algorithmic practices effectively (Bennett & Raab, 2006).

Existing regulations tend to prioritize data protection over the ethical implications of algorithms themselves. This oversight presents challenges in addressing algorithmic accountability and fairness, as current frameworks are ill-equipped to tackle the unique complexities posed by algorithms (Kitchin, 2016). A nuanced understanding of these challenges is essential for developing adaptive regulatory approaches that can keep pace with the evolving landscape of algorithmic technologies.

The opacity of algorithmic decision-making presents significant challenges for accountability and public trust. The intricate nature of algorithms often obscures the rationale behind decisions, making it difficult for users to understand how their data is being utilized. This lack of transparency fosters a climate of uncertainty, where individuals may be unaware of algorithmic biases that could impact their outcomes (Diakopoulos, 2016). As calls for greater transparency grow louder, the ethical implications of algorithmic decision-making demand urgent attention.

The black-box nature of algorithms raises ethical concerns regarding accountability in decision-making processes. When algorithmic decisions result in harm or erroneous outcomes, pinpointing liability becomes a complex issue. Should responsibility rest with the developers, the companies, or the algorithms themselves? This ambiguity complicates the establishment of accountability frameworks that address the ethical implications of algorithmic decision-making, highlighting the need for clear guidelines and regulations (O'Neil, 2016).

Moreover, the dynamic nature of algorithms complicates regulatory efforts, as these systems are constantly evolving through machine learning (ML) techniques. The ability of algorithms to adapt and improve over time poses challenges for regulators, who may struggle to keep pace with ongoing changes (Binns, 2018a, 2018b). To effectively address the challenges presented by algorithms, regulatory approaches must be adaptive and capable of accommodating future developments in algorithmic technologies.

Navigating the balance between innovation and public welfare is crucial in the discourse surrounding algorithm control. While algorithms have the potential to drive advancements in various fields, their unchecked deployment can lead to unintended consequences (Burrell, 2015). Ethical considerations must take center stage in discussions about algorithmic governance, emphasizing the importance of prioritizing human values and social well-being over mere profit motives.

Engaging diverse stakeholders in algorithm governance is vital to ensuring that a range of perspectives is considered in decision-making processes. Policymakers, technologists, ethicists, and affected communities must collaborate to create comprehensive frameworks that prioritize transparency, accountability, and inclusivity in algorithmic practices (Kitchin, 2016). By fostering an inclusive dialogue, we can better understand the complexities of algorithmic technologies and develop ethical guidelines that reflect the diverse needs of society.

Additionally, the development of ethical guidelines for algorithmic design and deployment is imperative to mitigate risks associated with algorithmic bias and discrimination. Organizations like the Partnership on AI and the Algorithmic Justice League advocate for ethical considerations in technology, emphasizing the importance of promoting fairness, accountability, and inclusivity in algorithmic systems (Partnership on AI, 2020). These initiatives serve as crucial stepping stones toward establishing a culture of ethical responsibility in the tech industry.

Advocacy for ethical standards in algorithmic design and deployment has gained momentum in recent years, driven by growing awareness of the societal implications of algorithmic technologies. Organizations and coalitions are actively working to establish guidelines that prioritize transparency, fairness, and accountability in algorithmic practices. Initiatives such as the Partnership on AI and the Algorithmic Justice League aim to promote ethical considerations and ensure that algorithmic technologies serve the broader public good (Algorithmic Justice League, 2021).

Collaboration between stakeholders is essential to foster a holistic approach to algorithmic ethics. This involves engaging technologists, ethicists, policymakers, and affected communities in meaningful dialogue. By incorporating diverse perspectives, we can develop comprehensive ethical frameworks that address the complexities of algorithmic decision-making and prioritize the well-being of individuals and communities (O'Neil, 2016).

Promoting ethical standards necessitates a shift in mindset within the tech industry, emphasizing responsibility alongside innovation. Companies must recognize their role as stewards of technology, accountable for the societal impacts of their algorithms. By fostering a culture of ethical awareness, organizations can proactively address potential biases and develop algorithms that promote equity and inclusivity (Binns, 2018a, 2018b). The future of algorithm governance lies in a proactive approach that anticipates and addresses the ethical implications of algorithmic technologies. As algorithms continue to permeate various aspects of society, fostering a culture of accountability and ethical consideration becomes paramount. This includes establishing clear guidelines for algorithmic transparency, ensuring diverse representation in algorithm development, and advocating for user rights in data privacy and decision-making processes (Diakopoulos, 2016).

Investing in education and media literacy initiatives will also play a vital role in empowering individuals to navigate the algorithmic landscape effectively. By equipping users with the skills to critically analyze algorithmic outputs, we can foster a more informed citizenry capable of engaging with technology in a responsible and ethical manner (Burrell, 2015). Schools, communities, and organizations should collaborate to develop educational programs that promote digital literacy and empower individuals to question and understand the implications of algorithms on their lives. Moreover, the establishment of independent oversight bodies can enhance accountability in algorithm governance. These entities could monitor algorithmic practices, conduct audits, and evaluate the societal impacts of algorithms. By providing a platform for public input and feedback, independent oversight can ensure that algorithmic technologies are developed and deployed in alignment with societal values.

QUANTUM ALGORITHMS

As AI autonomy raises ethical concerns, quantum computing introduces a new frontier, leveraging the strange yet powerful principles of quantum mechanics to redefine what's computationally possible. Unlike classical algorithms, quantum algorithms take advantage of quantum phenomena like superposition, entanglement, and quantum parallelism, allowing them to solve problems that classical computers can barely scratch the surface of. With quantum computing gaining traction, we're on the cusp of transformative changes across fields such as cryptography, optimization, AI, and scientific research.

At the forefront of these groundbreaking algorithms is Shor's Algorithm, which tackles integer factorization—a problem that exponentially increases in difficulty as the number size grows. Classical algorithms stumble here, with systems like RSA encryption relying on this very difficulty to secure digital communications. Shor's Algorithm, however, can factor large numbers exponentially faster than any known classical method (Shor, 1994), posing a monumental challenge to modern cryptography.

The implications are far-reaching. RSA encryption underpins everything from online banking to government communications, with large-scale integer factorization acting as a crucial barrier to cracking these systems. But if large-scale quantum computers become a reality, Shor's Algorithm could render RSA and other public-key cryptographic systems obsolete, opening the door to a cybersecurity crisis (Arute et al., 2019). In response, governments and corporations are investing in quantum-resistant cryptography, like lattice-based cryptography, which is believed to withstand quantum attacks (NIST, 2020). But transitioning to these new cryptographic standards is a slow and complex process, raising concerns about the interim vulnerabilities.

One compelling example of this looming threat came in 2019 when Google demonstrated a quantum computer that could perform calculations in mere moments that would take classical computers thousands of years. While still a proof of concept, this demonstration highlighted just how real these quantum advancements are and their potential to break current encryption methods (Arute et al., 2019).

On a more optimistic note, quantum cryptography techniques such as Quantum Key Distribution (QKD) are being developed to offer provably secure encryption. By exploiting the very principles of quantum mechanics, QKD promises a future where encryption keys are virtually impossible to intercept or replicate (Gisin et al., 2002). However, the practical implementation of QKD requires substantial upgrades to existing communications infrastructure, which will take time.

While Shor's Algorithm threatens cryptography, Grover's Algorithm presents a less immediate but still significant advantage in search optimization. Grover's Algorithm improves the time it takes to search through an unsorted database from $O(N)$ to $O(\sqrt{N})$, offering a quadratic speedup (Grover, 1996). This can revolutionize a variety of industries. For instance, Grover's search capabilities could enhance data mining tasks, improve pattern recognition in AI systems, and optimize supply chain management by drastically reducing the time required to find optimal routes (Cao et al., 2019). In drug discovery, quantum search techniques could help researchers quickly sift through massive molecular databases, accelerating the identification of promising drug candidates.

Though the speedup Grover offers is quadratic, its impact is far from small, particularly in fields that rely on analyzing vast datasets. Healthcare, for example, could benefit from faster diagnostics, with AI-powered quantum systems helping identify disease patterns by analyzing complex medical records and imaging scans.

Another fundamental quantum algorithm is the Quantum Fourier Transform (QFT), central to many quantum algorithms, including Shor's. The QFT allows certain operations, like integer factorization, to be performed exponentially faster than classical methods (Nielsen & Chuang, 2010). This could revolutionize scientific research, enabling more accurate simulations in fields like climate science and chemistry.

In climate science, quantum computers using QFT could simulate global climate models with unprecedented precision, aiding in better policy decisions to combat climate change (Preskill, 2018). In a related project, D-Wave Systems partnered with the University of Southern California to explore quantum annealers for climate modeling.

In chemistry, QFT could enable simulations of complex molecular interactions that were previously out of reach for classical supercomputers, advancing breakthroughs in drug discovery, materials science, and energy production (Babbush et al., 2018). IBM's Quantum Experience, for instance, is already using quantum simulations to study chemical reactions and material properties.

One of quantum computing's most impressive advantages is quantum parallelism, which allows quantum computers to explore multiple possibilities simultaneously. Classical systems process data sequentially, but quantum systems use superposition and entanglement to evaluate a vast number of potential outcomes all at once. This makes quantum algorithms incredibly efficient for problems with a large number of possible configurations, such as optimization.

In optimization, quantum parallelism is especially useful. For example, in finance, firms constantly search for optimal investment strategies. Quantum algorithms could help evaluate a wide range of scenarios simultaneously, drastically reducing time and improving decision-making. Volkswagen is already experimenting with quantum computing to optimize traffic flow and reduce congestion in cities, demonstrating practical applications of quantum algorithms in real-world settings (Volkswagen, 2019).

In biotechnology, quantum computing could simulate millions of potential drug interactions at once, greatly accelerating the drug discovery process (Cao et al., 2019). Likewise, in energy, quantum simulations could help design more efficient batteries and solar cells, supporting sustainable energy efforts.

As quantum computing matures, it promises to enhance ML and AI. Classical ML algorithms are limited by their reliance on classical data processing, but quantum computing opens the door to quantum machine learning (QML), which could revolutionize how AI systems function.

Quantum algorithms like the Quantum Support Vector Machine (QSVM) and Quantum Neural Networks (QNN) can drastically reduce the time needed to train

ML models by harnessing quantum parallelism (Broughton et al., 2021). These advancements could lead to faster and more accurate pattern recognition, classification, and optimization.

In healthcare, for instance, QML could drive personalized medicine by analyzing complex genomic and phenotypic data to create tailored treatment plans. One example is Zapata Computing's work on QML applications that improve drug discovery and diagnostics (Kumar et al., 2024).

In finance, QML could process massive datasets in real time, providing more accurate predictions of market behavior and enabling better investment decisions. Quantum-powered AI could also revolutionize fields like natural language processing, autonomous vehicles, and robotics.

Despite its immense potential, quantum computing is still in its early stages, with fully fault-tolerant quantum computers yet to be realized. To bridge the gap, researchers are developing hybrid algorithms that combine quantum and classical computing. Algorithms like the Variational Quantum Eigensolver (VQE) and the Quantum Approximate Optimization Algorithm (QAOA) utilize quantum processors for specific subproblems while relying on classical systems for others (Cerezo et al., 2021).

These hybrid approaches allow researchers to harness quantum power even with current hardware limitations. For example, in materials science, VQE is being used to simulate molecular properties, a task that is computationally intensive for classical systems alone (Peruzzo et al., 2014). A collaboration between IBM and MIT is exploring the use of VQE for material discovery.

Similarly, in logistics, QAOA is helping optimize complex routing problems by combining quantum and classical computations. D-Wave Systems has successfully applied hybrid approaches to solve logistical challenges, proving the practical applications of quantum computing in real-world business operations.

QUANTUM ALERT

Quantum algorithms are a double-edged sword, offering revolutionary potential while posing serious risks across various domains. They promise unprecedented advancements in fields ranging from drug discovery and materials science to optimization and AI. However, their most immediate and pressing threat lies in the realm of cybersecurity, where quantum algorithms could unravel the cryptographic fabric that secures our digital lives. Shor's Algorithm, in particular, threatens to dismantle widely used public-key cryptographic systems such as RSA encryption, which forms the backbone of internet security. If implemented on a sufficiently powerful quantum computer, Shor's Algorithm could factor large numbers exponentially faster than classical computers, rendering current encryption methods obsolete and potentially triggering a global cybersecurity crisis.

In the wrong hands, quantum computers could expose personal data, financial information, and even state secrets to unprecedented exploitation. Cybercriminals and malicious state actors would have the means to crack encryption codes that protect everything from personal emails and bank transactions to government communications and critical infrastructure. This scenario underscores the urgency of developing quantum-resistant cryptographic systems to safeguard digital security in a post-quantum world.

This isn't just a hypothetical danger—state actors are actively pursuing quantum supremacy as a cornerstone of cyber warfare strategies. Countries like China are investing heavily in quantum research, not only for technological and scientific leadership but also to gain a strategic advantage over global adversaries. China's establishment of the National Laboratory for Quantum Information Sciences exemplifies its commitment to becoming a quantum superpower, aiming to dominate both economic and military spheres through advanced quantum capabilities. This arms race raises crucial ethical questions about the deployment of quantum technologies for espionage, surveillance, and other covert activities that could destabilize international relations and infringe on individual rights.

Moreover, as quantum systems grow more sophisticated, the complexity and opacity of their algorithms could lead to a new frontier of algorithmic accountability issues. Quantum algorithms are fundamentally different from classical ones, often operating in ways that are difficult to interpret or explain even by experts. This lack of transparency poses significant challenges in ensuring that these algorithms behave as intended and do not perpetuate biases or errors.

Imagine a quantum algorithm designed to predict criminal behavior, a tool that could be tempting for law enforcement agencies seeking to leverage cutting-edge technology. However, if the decision-making process of this algorithm is too complex to be understood or scrutinized, it could embed and perpetuate biases that further entrench social inequalities. Such a scenario raises profound ethical concerns about fairness, transparency, and the potential for discrimination. In high-stakes areas like criminal justice, finance, and national security, the opacity of quantum algorithms could erode public trust and exacerbate existing disparities, leading to outcomes that are neither just nor equitable (O'Neil, 2016).

The duality of quantum algorithms—offering both immense potential and significant risks—demands a balanced approach. Policymakers, technologists, and ethicists must collaborate to ensure that the deployment of quantum technologies serves the public good without compromising security, privacy, or ethical standards. This involves not only advancing quantum-resistant cryptography but also establishing robust frameworks for algorithmic transparency, accountability, and fairness in this quantum era.

DATA PATROL

Existing laws and regulations have been sluggish in adapting to the rapid surge of AI and quantum technologies, a lag that is increasingly problematic in our fast-evolving digital landscape. In some regions, AI falls under general data protection frameworks like the GDPR in Europe, but these often miss the mark when it comes to addressing the nuanced challenges posed by autonomous systems. Liability laws, which vary by country, further complicate the global handling of AI-related harm. The European Union's AI Act stands out as a bold attempt to regulate AI, especially in high-risk domains such as facial recognition and biometric surveillance (European Commission, 2021). Yet, many other regions are scrambling to catch up, and even the most progressive regulations struggle to keep pace with relentless innovation. Current laws tend to zero in on data privacy and security, glossing over the ethical and legal conundrums intrinsic to autonomous decision-making.

To go through this brave new world, a host of practical solutions have been floated. First up: international standards for ethical AI development, creating a baseline of safety, fairness, and transparency. Independent algorithmic audits could vet AI systems for biases and ethical compliance before they are unleashed at scale. Think of ethical AI certification programs as the new organic label for tech—a stamp of approval that assures users of the systems' adherence to ethical standards. A fitting example is the Algorithm Transparency Standard from the Institute of Electrical and Electronics Engineers (IEEE), which aims to shine a light on AI algorithms and their applications (IEEE, 2019). Then there's the idea of "kill switches" in AI systems, ensuring human oversight in critical scenarios. These emergency shut-off mechanisms, already used in self-driving cars to prevent accidents, could be expanded to other AI systems to maintain control over autonomous technologies (Borenstein et al., 2019).

But regulation isn't just a top-down affair; it needs grassroots input. Public engagement is key, given AI's profound societal impact. AI literacy programs could empower citizens to grasp how AI shapes their lives, equipping them to weigh in on policy debates. Initiatives like public consultations and citizen juries, where everyday folks deliberate on AI issues, could yield valuable insights into balancing innovation with ethical considerations.

International cooperation is another must, given AI's borderless nature. A system developed in one corner of the globe can be deployed worldwide, necessitating cohesive regulatory frameworks. Bodies like the United Nations or the OECD could spearhead the creation of global standards for AI ethics, ensuring cross-border consistency. National governments, on the other hand, could establish specialized AI ethics boards to oversee the local development and application of AI technologies. These boards should be diverse, comprising technical experts, ethicists, legal scholars, and civil society representatives to ensure a well-rounded approach to regulation.

The integration of quantum algorithms into AI systems brings both tantalizing opportunities and hefty challenges. Quantum computing could supercharge AI capabilities, solving complex problems at previously unattainable speeds. Algorithms like Shor's, which can factor large numbers exponentially faster than classical methods, exemplify this potential (Nielsen & Chuang, 2010). Such advancements could revolutionize fields like cryptography, optimization, and ML. Yet, the rise of quantum algorithms also poses serious ethical and security concerns that demand regulatory foresight.

One major challenge is the potential threat to existing encryption methods. Classical encryption systems, essential for safeguarding sensitive information, could crumble under quantum computing's prowess. For instance, RSA and ECC encryption standards might become obsolete, leaving data vulnerable to quantum attacks. This scenario underscores the urgent need for quantum-safe cryptographic standards, a task for which global cooperation is paramount (NIST, 2020).

In this new era, the collaboration among tech companies, governments, and civil society is more crucial than ever. Companies must embed ethical considerations into their development processes, establishing internal ethics committees and conducting impact assessments. Giants like Microsoft and Google have already started implementing ethical guidelines, emphasizing principles like transparency and fairness (Microsoft, 2018; Google, 2020). Such initiatives are crucial for ensuring that quantum-enhanced AI systems align with societal values.

Governments play a crucial role in crafting adaptive regulatory frameworks that can evolve alongside AI and quantum technologies. Overly rigid or outdated regulations may stifle innovation or fail to address emerging risks. Regulatory bodies must engage with technology experts, ethicists, and industry stakeholders to create dynamic frameworks that can adapt to rapid advancements in quantum computing and AI. This includes regular reviews of existing regulations and the establishment of flexible guidelines that can accommodate new technologies as they arise.

Civil society organizations can serve as essential watchdogs, ensuring that AI systems enhanced by quantum algorithms remain transparent, fair, and accountable. These organizations can advocate for the rights of individuals and communities affected by AI technologies, ensuring that their voices are included in discussions about ethical and regulatory frameworks. By promoting public awareness and understanding of quantum computing and its implications for AI, civil society can contribute to informed discourse around these technologies.

COD-A

To envision a future where technology genuinely serves the greater good, we must first grapple with the power dynamics embedded in algorithmic governance. By fostering meaningful collaboration among stakeholders and championing robust ethical standards, we can sculpt a digital landscape that not only upholds human values but also protects individual rights and promotes social cohesion. The decisions we make today about algorithmic design and deployment will ripple through society, influencing norms, behaviors, and the overall well-being of humanity for generations to come.

Standing at the crossroads of technological advancement and ethical responsibility, the urgency for critical engagement with algorithms has never been more acute. A future where algorithms—including quantum algorithms, AI, robotics, biotechnology, and data analytics—are firmly rooted in ethical principles can elevate society rather than fracture it, ultimately shaping the legacy of the algorithmic age (O'Neil, 2016; Russell & Norvig, 2020). This vision requires not just vigilance and advocacy but also an unwavering dedication to cultivating a digital world marked by inclusivity, equity, and justice (Binns, 2018a, 2018b; Cath, 2018).

As AI, quantum computing, robotics, and other emerging technologies redefine the boundaries of intelligence and consciousness, we must also reevaluate humanity's place within the broader ecosystem of life. The increasing integration of these advancements into every facet of human activity—from governance to art—raises profound questions about the future of humanity and its relationship with the planet (Harari, 2018). In the next and final chapter, we will explore the ecological dimensions of these technologies, examining their environmental impact, their role in degrading but also preserving biodiversity, and how they may reshape our understanding of our role as "humans" in the Anthropocene.

10 The Byte of the Earth

The term "human" first appeared in the mid-13th century, derived from the Middle French word *humain*, meaning "pertaining to man." This etymological trail reveals a deep-seated connection between humanity and the earth, hinting at an identity historically intertwined with nature. In Middle French, *humain* surfaced in various literary works, notably the *Roman de la Rose* (1230), a foundational text exploring love and human nature (Benson, 1982). In these early contexts, *humain* conveyed more than biological distinction; it embodied the intricate dance between humans and the environment. The essence of *being human* resonated with an awareness of our earthly roots, underscoring our intrinsic bond with the land that sustains us.

The journey of *humain* traces back to the Latin *humanus*, linking two roots: *homo* (man) and *humus* (earth). *Humus*, meaning "soil" or "ground," accentuates humanity's physical and existential grounding in the earth (Rosenberg, 2009). This duality implies that to be human transcends intellect or culture, rooting us in the very soil we emerge from. The Romans perceived this connection, not merely in a literal sense but as a cultural ethos, where being human encapsulated physical existence and a moral duty toward the earth.

The Romans introduced *humanitas*, a concept enveloping compassion, education, and civic duty, contrasting the civilized *homo humanus* with the uncultured *homo barbarus* (Nussbaum, 1997). This dichotomy linked civilization to land cultivation, as Aristotle highlighted in *Politics*, framing the civilized Greeks against non-Greeks, often perceived as disconnected from the cultivated earth (Aristotle, 1996). Thus, one's humanity was measured by their ties to community and environment.

Cicero, a pivotal figure in these discussions, associated humanity with respect for nature and the cosmos, positing that true *humanitas* emerged from harmonious interactions with the land and each other (Cicero, 2005). The Roman landscape mirrored societal values, with its agrarian roots reinforcing their identity as earthbound humans.

Expanding this discourse, Quintilian built upon Cicero's ideals, emphasizing education's role in cultivating individuals who embodied the moral virtues tied to *humanitas*. In *Institutio Oratoria*, Quintilian argued that the pinnacle of eloquence and knowledge should steer individuals toward ethical living, deeply connected to humanity and the natural world (Quintilian, 1920). For Quintilian, nature was not just a backdrop but integral to moral and intellectual development, reinforcing ethical conduct tied to the land.

This intellectual trajectory continued with Latin playwrights like Plautus and Terentius, who explored human relationships and societal norms through the lens of *humanus*. Plautus's *Asinaria* metaphorically linked humanity's predatory nature to disconnection from place (Plautus, 2006). Terentius's famous line, "Homo sum: humani nihil a me alienum puto," translated as "I am a man; nothing human is alien to me," emphasized shared humanity and our connection to the earth (Terentius, 1976).

DOI: 10.1201/9781003624813-11

As *humanitas* evolved, it influenced ethical and moral responsibility discussions, leading to the Renaissance's revival of these ideals. This period urged a reevaluation of human ambition versus environmental sustainability, promoting stewardship of the earth as central to humanity's role (McKibben, 2010). The Renaissance fostered a holistic respect for nature, recognizing the planet's health as vital to its inhabitants' well-being.

Our current trajectory in relating to the planet is a far cry from what it could be. In an age where pollution, environmental degradation, and climate change dominate headlines, it's clear that humanity has become disturbingly disconnected from the natural world. The relentless drive for profit has fueled the reckless exploitation of Earth's resources, leading to ecological disasters and displacing countless vulnerable communities. Climate change is not just an abstract future threat; it's a present-day crisis manifesting in extreme weather, rising seas, and a catastrophic loss of biodiversity.

Consider the Flint water crisis in Michigan—a chilling example of toxic environments' human cost. When lead contaminated the city's water supply, it wasn't just the pipes that were corroded but the trust and health of its residents, who suffered a cascade of illnesses and heightened anxiety. Similarly, the 2010 Deepwater Horizon oil spill in the Gulf of Mexico spilled about 4.9 million barrels of oil into the ocean, devastating marine life and the livelihoods of coastal communities dependent on fishing and tourism. These events aren't isolated mishaps; they are glaring symptoms of a society that values material consumption over ecological harmony.

But the damage isn't confined to dramatic disasters. The air we breathe in many urban areas is now a toxic mix, contributing to an estimated 4.2 million premature deaths annually, with low-income communities bearing the brunt (World Health Organization, 2024a, 2024b). This insidious killer doesn't just clog lungs—it fuels anxiety and depression as people grapple with the grim reality of living in polluted environments.

Meanwhile, the global waste crisis looms large. Consumer culture has given rise to mountains of plastic and electronic waste, much of which is non-biodegradable and toxic. The Great Pacific Garbage Patch, a vast floating island of trash, is a haunting emblem of our disconnection from nature. It not only threatens marine life but also disrupts ecosystems and food chains, pushing species to the brink (Lebreton et al., 2018).

The impacts of climate change stretch further still. In southern Africa, severe droughts have forced countries like Zimbabwe and Namibia to cull wild elephants and other animals to stave off starvation in human populations. Zimbabwe plans to slaughter 200 elephants, while Namibia is culling over 700 wild animals, including 83 elephants, to provide protein-rich meat to hungry communities (Associated Press, 2024). This grim calculus underscores the desperate balancing act between conservation and survival, revealing how climate change can force nations into heartbreaking decisions that erode both biodiversity and cultural heritage.

As we distance ourselves from the earth, we lose sight of our intrinsic connection to it. This growing detachment breeds exploitation rather than stewardship, obscuring our responsibility to protect the complex web of life that sustains us. Prioritizing convenience and short-term gains over long-term sustainability jeopardizes not just the planet but the legacy we leave for future generations.

As artificial intelligence (AI) continues its rapid transformation of industries, its environmental toll is becoming increasingly hard to ignore. The ecological footprint of AI stretches beyond the immediate impact of energy consumption in massive data centers; it includes the extraction of raw materials and the lifecycle of the technology that powers AI systems (Strubell et al., 2019). These hidden environmental costs, often relegated to the sidelines, have profound implications for our planet's future.

AI hardware, from servers and GPUs to sensors, relies heavily on rare earth metals like lithium, cobalt, and nickel. The extraction of these minerals, often in developing nations, has caused severe environmental damage—deforestation, habitat loss, and water contamination are common consequences of mining operations (Baldé et al., 2015). In places like Chile's Atacama Desert, lithium mining has worsened water shortages, setting the stage for conflict between local communities, agriculture, and the mining industry.

The manufacturing of AI components also generates a staggering amount of e-waste. As hardware becomes obsolete, toxic materials leach from discarded electronics, contaminating the soil and water in landfills (Baldé et al., 2015). This waste stream reflects a broader environmental crisis, where the pursuit of technological progress increasingly distances us from the ecosystems that sustain life.

Once operational, AI systems require energy-hungry data centers to store and process data, contributing significantly to global energy consumption. In fact, the tech industry is responsible for about 2–3% of global carbon emissions—comparable to the emissions of the aviation sector. These data centers must run 24/7, consuming vast amounts of energy not only for processing but also for cooling. The demand for computational power has surged with the advent of large-scale machine learning models, such as natural language processing and image recognition algorithms, further taxing energy grids.

Take, for instance, GPT-3, the language model created by OpenAI. The energy required to train GPT-3 amounted to hundreds of megawatt-hours, a stark reminder of the environmental costs of developing cutting-edge AI systems (Amodei et al., 2016). This highlights the urgent need for a more sustainable approach to AI, one that emphasizes energy-efficient algorithms, the use of renewable energy in data centers, and innovations in hardware design that reduce overall power consumption.

A recent study sheds light on the considerable water consumption involved in running AI models like ChatGPT. According to research conducted by the Washington Post in partnership with the University of California, every 100-word email generated by AI consumes roughly the equivalent of a bottle of water. This is due to the cooling systems that are integral to maintaining data centers (Heath, 2023). These findings underline the need for a holistic, sustainability-minded approach to the environmental costs of emerging technologies.

This reality forces us to reconsider what it means to be human in an age dominated by technological progress. It is a stark reminder that our existence is inextricably linked to the earth that sustains us. While we push the boundaries of innovation, we must also cultivate a deep sense of responsibility to our environment, prioritizing sustainability in every aspect of our technological development. The urgency of this change requires us to reevaluate our values, shifting toward a worldview that recognizes the interconnectedness of all life.

When facing these challenges, we can turn to ancient philosophical traditions for guidance. The Stoics, for example, championed a balanced relationship between humanity and nature, teaching that true wisdom lies in understanding our place within the natural order (Hadot, 1995). By adopting principles of moderation, humility, and respect for the environment, we can begin to heal the rift between humanity and the planet.

ECO-AI

While AI and quantum computing pose significant environmental costs, they also present remarkable opportunities to tackle pressing ecological challenges, particularly in biodiversity conservation. AI-driven technologies have the potential to revolutionize how we protect endangered species, manage natural resources, and respond to environmental threats.

Globally, AI is making notable contributions to wildlife conservation. For instance, AI-powered drones and camera traps equipped with machine learning algorithms monitor animal populations, track migration patterns, and detect poachers in real time. A prime example is the Wildlife Conservation Society's use of AI technologies to protect endangered elephants in Africa. By analyzing GPS tracking data, conservationists can predict poaching activities and conduct targeted patrols, significantly reducing illegal killings for ivory (WCS, 2020).

In addition to terrestrial efforts, AI is vital in ocean conservation. Researchers utilize AI-driven analyses of underwater audio recordings to monitor marine biodiversity. For instance, Ocean Networks Canada employs AI to identify the sounds of various marine species, such as whales and dolphins, aiding in assessing ecosystem health and detecting illegal fishing (ONC, 2021). This technology not only protects endangered marine life but also promotes sustainable fishing practices essential for long-term viability.

Moreover, AI plays a crucial role in reforestation initiatives. The World Resources Institute (WRI) uses machine learning models to analyze satellite imagery, identifying areas for reforestation and determining optimal tree species based on local soil and climate conditions. Projects like Ecosia, a search engine that funds tree-planting initiatives, leverage AI to enhance the impact of their efforts. By pinpointing suitable locations and species, AI fosters healthier ecosystems and combats climate change (Omdena, 2022).

AI also enhances sustainable agriculture through precision farming techniques. Smart sensors, drones, and AI algorithms monitor crop health, soil conditions, and weather patterns, optimizing irrigation, pesticide use, and fertilization. For example, IBM's Watson Decision Platform for Agriculture provides farmers with insights from weather data and crop health metrics, enabling them to maximize yield while minimizing environmental impact (Content CES, 2021). This approach conserves resources and reduces chemical runoff into waterways, promoting healthier ecosystems.

Automated irrigation systems further exemplify AI's impact on sustainability. Companies like CropX utilize soil sensors and AI to determine optimal water amounts for crops, significantly reducing water waste by up to 50%. This ensures crops receive adequate hydration while conserving precious water resources, particularly in drought-prone areas. Additionally, quantum computing holds promise for applications that can help save the planet. Quantum algorithms optimize complex systems like energy grids and transportation networks, leading to more efficient resource use. Companies such as Google and IBM explore quantum computing's potential to enhance renewable energy systems. Google's Quantum AI team investigates how quantum computing can optimize solar energy forecasts and improve energy storage solutions. Furthermore, researchers at the California Institute of Technology employ quantum algorithms to analyze climate models effectively. By processing vast amounts of data at unprecedented speeds, quantum computers enhance our understanding of climate change impacts and develop more effective mitigation strategies. These advancements could significantly reduce carbon emissions and promote a sustainable energy future.

AI improves climate modeling and predictions, aiding policymakers in making informed decisions. Microsoft's AI for Earth initiative utilizes machine learning algorithms to analyze climate data, predict extreme weather events, assess climate change impacts, and model mitigation strategies (Microsoft, 2025). This proactive approach enables governments and organizations to develop effective disaster response plans and implement policies addressing the root causes of climate change. One notable example is ClimaCell, which uses AI to provide hyper-local weather forecasts and alerts. By analyzing real-time data from various sources, ClimaCell helps communities prepare for severe weather events, ultimately reducing vulnerability and enhancing resilience against climate-related disasters.

AI technologies also contribute to building smarter cities that prioritize sustainability. For instance, Barcelona has implemented a smart traffic management system using AI to analyze traffic patterns and adjust signals in real time, reducing congestion and emissions. Additionally, AI optimizes energy usage in buildings, with the Green Building Council reporting that AI-driven systems can reduce energy consumption by up to 30% by learning usage patterns and adjusting heating, cooling, and lighting systems accordingly. The Nest Learning Thermostat exemplifies this application, optimizing home energy use based on learned user behavior, leading to significant energy savings.

The Internet of Things (IoT), combined with AI, plays a crucial role in waste management. Smart waste bins equipped with sensors monitor fill levels and optimize collection routes, reducing fuel consumption and emissions associated with garbage collection. Rubicon, a waste management technology company, uses AI to analyze waste data and provide businesses with actionable insights to reduce waste and increase recycling rates (SWANA, 2021). Cities implementing these technologies can transition toward a circular economy that minimizes waste and conserves resources, as seen in Rome, Italy, where a Smart Waste Management System uses IoT sensors to monitor waste levels in public bins, allowing for efficient waste collection schedules (Boresta et al., 2024).

However, while the promise of AI and quantum computing is vast, ethical concerns must be addressed to ensure sustainable practices. The deployment of AI technologies must respect privacy and local communities. Conservation efforts should actively engage local populations to ensure that AI solutions align with their needs and rights. The UNESCO framework on AI emphasizes the importance of ethical considerations in deploying AI across sectors for all 194 members, including environmental conservation (UNESCO, 2023).

BEYOND ANTHROPOCENE

The Anthropocene—a term that signifies an era of unprecedented human influence on the Earth's ecosystems—challenges humanity to rethink its relationship with nature. As the dominant force shaping the planet's climate and biodiversity, human activity has triggered phenomena like climate change, habitat destruction, and species extinction (Steffen et al., 2011). The evidence is irrefutable: rising global temperatures, ocean acidification, and the intensification of extreme weather events are all linked directly to human actions (Ripple et al., 2024). The relentless consumption of natural resources, alongside deforestation and pollution, has brought many ecosystems to the brink of collapse, demanding urgent, creative solutions.

Enter AI and quantum computing—not just byproducts of the Anthropocene, but also potential lifelines in addressing the environmental crises they have helped exacerbate. These technologies have already shown promise in enhancing climate science: AI models are refining our understanding of climate trends, forecasting extreme weather events, and suggesting new strategies to counteract climate impacts. AI has modeled sea-level rise and predicted polar ice melting, providing policymakers with invaluable data for informed decision-making (Huang et al., 2024). AI's environmental monitoring capabilities are equally impressive: algorithms track deforestation rates, assess coral reef health, and even predict wildfires. In California, AI-driven forecasts have allowed authorities to take timely preventative action, saving lives and protecting property.

However, the Anthropocene compels a more profound reflection on philosophical and ethical dilemmas surrounding our treatment of nonhuman life. While AI offers tools to alleviate environmental damage, its development must be scrutinized through an ethical lens that challenges the anthropocentric assumptions embedded in modern technology and governance. Gabriel Giorgi's *The Animal's Turn: Thinking with Animals in the Anthropocene* critiques the long-standing divide between humans and nonhuman animals—a divide that contributes to many of the ecological crises we face today. Giorgi's work aligns with the broader "Animal Turn" in critical theory, a movement that seeks to place animals at the heart of ethical, philosophical, and political discourse (Giorgi, 2020).

The "Animal Turn" prompts us to reconsider the ethical boundaries separating humans from nonhuman animals. Thinkers like Jacques Derrida and Giorgio Agamben argue that the Anthropocene reveals the untenability of these divisions. In *The Animal That Therefore I Am*, Derrida dismantles the Cartesian view of animals as mindless machines—a view that has justified their exploitation for centuries. Derrida's concept of *animot* challenges the singular category of "the animal" and instead emphasizes the diversity and complexity of nonhuman life, urging us to acknowledge the agency and subjectivity of animals (Derrida, 2002).

Similarly, Agamben's *The Open: Man and Animal* interrogates the exclusion of animals from ethical and political spheres, critiquing the concept of "bare life"—a life subject to control, violence, and exploitation without moral consequence (Agamben, 2004). In the Anthropocene, both humans and animals increasingly face the reduction to bare life as ecosystems collapse and the commodification of nature accelerates. Agamben and Derrida's insights point to the necessity of moving beyond anthropocentrism, advocating for a more inclusive ethics that considers the rights and interests of all life forms.

Affect theory plays a pivotal role in reshaping our understanding of human-animal relations in the Anthropocene. Thinkers like Donna Haraway argue that animals are not mere victims of human actions, but active participants in shaping human history, culture, and ethics. Haraway's concept of *companion species* stresses the co-evolutionary relationship between humans and animals, positioning them within an interconnected web of affect and influence (Haraway, 2008). Recognizing this interdependence is crucial for developing ethical frameworks that move beyond human-centered concerns to embrace nonhuman animals as equals.

Affect theory also challenges the rationalist and mechanistic views of animals that have dominated modernity. Brian Massumi's work on affect reveals how emotions, sensations, and energies flow between human and nonhuman bodies, eroding the rigid boundaries between them (Massumi, 2015). This view helps us see animals as beings with their own subjective experiences, rather than merely as resources or tools. In the context of the Anthropocene, such an approach calls for ethical and political frameworks that honor the shared vulnerability of all living beings in the face of ecological degradation.

AI's integration with animal ethics is evident in contemporary conservation efforts. For instance, AI is being used to monitor endangered species, track migration patterns, and study animal behavior in unprecedented detail. The Snow Leopard Trust, for example, employs AI to analyze images from camera traps, identifying individual snow leopards and tracking their populations in remote areas of Central Asia. This approach not only aids in conservation but also challenges traditional perceptions of animals as passive subjects, recognizing them as unique individuals. Similarly, AI is transforming coral reef monitoring. As climate change and pollution threaten coral ecosystems, AI algorithms are employed to analyze underwater images, assess coral health, detect bleaching events, and predict future changes. These efforts go beyond preserving ecosystems for human benefit; they recognize the intrinsic value of nonhuman life and the moral obligation to protect it.

These examples illustrate how AI can align with the ethical insights of the "Animal Turn," yet they also raise critical questions about the role of technology in managing nature. As AI becomes increasingly involved in conservation, we must ensure that these tools don't reinforce the anthropocentric frameworks that have driven ecological crises. Instead, AI should help foster a more respectful, inclusive relationship with animals and ecosystems, one that honors their rights and intrinsic value.

The ethical dilemmas posed by AI in the Anthropocene extend beyond efficiency and effectiveness; they prompt fundamental questions about who controls these technologies and for what purposes. As Luciano Floridi asserts, AI's role in environmental governance must be grounded in fairness, transparency, and accountability, particularly in regions most vulnerable to climate change (Floridi, 2014). Without careful implementation, AI could exacerbate existing inequalities, underscoring the importance of equitable policies.

Furthermore, the philosophical frameworks of Derrida, Agamben, and other contemporary thinkers remind us that AI should not simply be a tool to mitigate human damage to the environment. Rather, it should contribute to a broader ethical project that reimagines our responsibilities toward nonhuman life, creating technologies that respect the agency and affect of animals and ecosystems, rather than treating them as mere resources to be exploited or problems to be solved.

WHEN MACHINES LISTEN

As AI becomes increasingly integrated into our efforts to tackle the challenges of the Anthropocene, it raises ethical questions that demand our immediate attention. Who holds the reins over the data and models that steer environmental decision-making? How can we guarantee that AI is used equitably, especially in regions most vulnerable to climate change? These questions stress the pressing need for global governance frameworks that ensure AI is deployed responsibly and transparently.

The role of AI in environmental governance transcends technical applications; it also touches on fundamental issues of power, justice, and representation. Who benefits from these AI systems, and at what cost? Ensuring that marginalized communities, often the hardest hit by climate change, are included in the design and implementation of AI solutions is crucial. This inclusivity is not just vital to mitigate potential biases in datasets but also to guarantee that solutions are both relevant and effective in the contexts where they are applied.

Moreover, collaboration between technologists, ethicists, and local communities can create AI systems that are not only efficient but also aligned with the values and needs of the people they serve. For example, in areas prone to climate-induced natural disasters, local knowledge can enhance AI algorithms designed to predict floods or droughts, improving their accuracy and reliability. This collaboration transforms communities from passive recipients of technology into active participants in shaping their futures.

Earlier, we referred to the origins of the term "human" and its historical usage, particularly in the writings of Marcus Fabius Quintilianus (c. 35–c. 100 AD). His question, "Do we call a man human merely because he is born of the earth?"—translated as *Etiamne hominem appellari, quia sit humo natus*—invites us to reflect on the symbolic boundaries of our shared existence (Quintilian, 1920). This notion of "earthly beings" has been defined in opposition to the divine, the bestial, and the barbaric, highlighting the historical and cultural forces that shape our understanding of humanity.

It's essential to recognize that the Latin concept of *humanitas* was defined not just by clear-cut divisions—such as the contrast between the divine and the beastly—but also by subtler exclusions. Groups such as women, children, and slaves were systematically excluded from the conversation about humanity's scope. This opposition between *Homo humanus* and *homo barbarus* was not confined to ancient Rome; it has historically marginalized entire communities, establishing hierarchical systems of worth and value. Similarly, the historic division between humans and animals may resurface in contemporary debates on clones, humanoids, and AI entities.

The evolution of this concept signifies a shift from traditional Roman values, such as *mos maiorum*, which emphasized family lineage, civic duty, and collectivism. However, the elitism, sexism, and classism embedded in these values must be acknowledged. Quintilian's query reminds us that the discussion surrounding humanity has often been narrowly confined, posing significant questions about inclusivity in today's dialogues on AI and environmental governance.

For instance, AI systems deployed in agriculture frequently exclude the perspectives of smallholder farmers, who often lack the resources of larger agribusinesses. By overlooking these voices, AI can exacerbate existing inequalities rather than mitigate them. If AI is to contribute to environmental sustainability, it is imperative that we actively seek out and incorporate the insights of those most affected by climate change and environmental degradation.

Additionally, the integration of AI into environmental monitoring systems must account for local ecological knowledge and cultural practices. Indigenous communities, for example, often hold valuable traditional ecological knowledge that can enhance AI models for biodiversity and land use (Berkes, 2018). Incorporating such knowledge not only improves AI's accuracy but also underscores the importance of diverse perspectives in environmental stewardship.

Throughout this chapter, we have presented examples that challenge historical divisions within humanity and between humans and other entities. Participatory AI design—such as community-led data collection initiatives and local advisory boards—can bridge these divides. Projects like "Data for Good," which use data science for social impact, demonstrate how technology can be repurposed to benefit underrepresented communities. By ensuring that AI is developed with input from those it affects, we can achieve more equitable outcomes.

We have also explored hybrids—ancient and modern—that challenge the traditional concept of humanity as a singular, immutable entity. These hybrids encompass a broad spectrum of beings, from mythical creatures to genetically modified organisms (GMOs) and AI entities. They challenge dichotomies such as human/non-human, rational/irrational, and civilized/barbaric (Latour, 2005). Acknowledging the fluidity of these boundaries opens the door to discussions about how technology can better reflect the complexity of existence.

In ancient cultures, hybrid beings often symbolized the interconnectedness of life. For example, centaurs in Greek mythology represented a blend of human intellect and animal instinct, reflecting the complexities of human nature. Similarly, deities like Ganesha in Hinduism, with his elephant head and human body, symbolize the idea that the divine transcends rigid categories. These hybrids demonstrate that human experience cannot be neatly categorized; it encompasses a spectrum that defies simplistic binaries.

As we move deeper into discussions about the intersection of technology, life, and ethics, the concept of hybrids continues to evolve. Today, genetic engineering and biotechnology challenge traditional understandings of what constitutes life. GMOs, for example, force us to confront questions about what it means to be "natural" and challenge the notions of ownership, environmental impact, and the rights of living beings. As we manipulate life forms for our benefit, we must consider the moral consequences of our actions and the responsibilities they entail.

AI represents another modern hybrid that complicates our traditional definitions of humanity. AI systems, capable of learning, adapting, and exhibiting behaviors that mimic human decision-making, push the boundaries of our assumptions about consciousness, agency, and emotionality (Huriye, 2023). As AI becomes more integrated into our lives, we must wrestle with the question of whether these systems can embody characteristics we traditionally associate with humanity, such as empathy or creativity. This blurring of lines calls for a reassessment of our relationships with non-human entities and a reconsideration of the possibilities for coexistence and collaboration.

By recognizing the fluidity of these boundaries, we can approach technology not as an external force designed to serve human needs but as an extension of our collective intelligence—a tool capable of amplifying diverse voices and perspectives. Participatory design in technology development encourages collaboration between technologists and communities, ensuring that these systems are not just functional but also culturally and ecologically sensitive.

Indigenous practices, which often merge human and non-human elements, provide an insightful perspective on our interconnectedness with the natural world. Rooted in Native American spirituality, these traditions emphasize the cyclical nature of life, where humans, animals, plants, and even inanimate elements of nature are seen as part of a larger, interconnected tapestry (Deloria, 2003). Each entity holds a distinct purpose, contributing to the dynamic relationship between the Earth and its inhabitants. This holistic view fosters a deep sense of responsibility toward nature, encouraging individuals to reflect on the consequences of their actions on the environment and their communities.

Central to this Indigenous perspective is a commitment to sustainability and ecological balance. Passed down through generations, traditional ecological knowledge informs practices such as agriculture, hunting, and gathering, nurturing the stewardship necessary for the survival of both the land and its people. Elders' stories, which underscore the delicate balance of ecosystems, instill respect for wildlife and natural resources, shaping younger generations into responsible caretakers of the Earth. In this worldview, being human means acting in harmony with the natural world, fostering an interconnectedness that transcends human existence.

In the context of climate change, these Indigenous perspectives call for a holistic approach to environmental governance. Recognizing the interdependence of all life forms encourages policies that respect traditional practices and integrate local ecological knowledge. By incorporating these worldviews into decision-making, we can craft more effective and sustainable solutions to the environmental crises we face.

AI has the potential to benefit significantly from this perspective. By integrating traditional ecological knowledge into AI-driven models, we can create more innovative and ecologically sound strategies for resource management and conservation. Many Indigenous communities are already using tools like geographic information systems (GIS) to document their traditional lands and resources, strengthening their advocacy for environmental protection. These technologies not only empower these communities but also enhance their ability to protect the environment through responsible stewardship.

The fusion of AI with Indigenous knowledge could lead to the development of more culturally relevant and ecologically grounded technological solutions. By prioritizing the input of those most affected by environmental policies, we ensure that technological advancements are aligned with the values and needs of diverse communities. This approach not only increases the effectiveness of AI applications but also fosters a sense of inclusivity and respect for marginalized groups.

Furthermore, Indigenous perspectives often emphasize mindfulness and presence— qualities that can be crucial in the development of AI technologies that foster ecological awareness. Mindfulness apps and virtual reality experiences inspired by these philosophies could enhance individuals' connection to the natural world, cultivating a deeper understanding of our role within the broader ecosystem. Through heightened awareness of our surroundings, we can develop a greater appreciation for the interconnectedness of all life, promoting sustainable and ethical choices in our interactions with technology.

As we conclude in the Epilogue, we will continue to explore these ideas, contemplating posthumanism and the potential end of the human era. This reflection forces us to consider not just the future of humanity, but the future of all life on Earth. By challenging traditional boundaries, we can imagine a more inclusive future—one in which technology, ethics, and environmental stewardship converge to create systems that reflect a shared humanity, embracing diversity in all its forms.

Epilogue
Ctrl+Alt+TheEnd

Throughout these pages, I have tiptoed around the central question that lingers beneath every chapter: *Have we ever truly been human?* And if so, who among us could be conceived as fully human in the way the original term was intended? The answer is uncomfortable, for history shows that our conception of "humanity" has never been inclusive, never universal. The idea of what it means to be human has always been shaped by exclusion—defined by those who hold power, and denied to those who do not fit neatly into these narrow definitions.

Dehumanization is not a new phenomenon; it has been a defining feature of human history. From colonial conquests to modern warfare, dehumanization has provided the moral and psychological scaffolding that allows individuals and entire societies to justify violence, exploitation, and exclusion. It is the underlying justification behind centuries of entrenched prejudice. We see it today in the way Israeli rhetoric sometimes frames Palestinians or how the Nazis spoke of Jews—not merely as enemies but as subhuman. These are not isolated examples; they are iterations of an ancient logic that reduces certain groups to mere symbols of threat or danger.

What makes dehumanization even more insidious today is how modern technology amplifies and normalizes these tendencies. Through digital media, entire populations can be dehumanized at alarming speed, unchecked by the traditional gatekeepers of truth. Hate speech, propaganda, and conspiracy theories spread through social media, normalizing dehumanizing narratives that ultimately fuel real-world violence. The impact is no longer limited to speeches or manifestos but plays out in our daily interactions with the bombardment of images, memes, and disinformation (Zuboff, 2019).

This digital battleground is not just a new medium for old hatreds; it represents a systemic shift in how we relate to one another as human beings. In our fragmented, algorithm-driven societies, empathy is often a casualty. Algorithms prioritize outrage, and media ecosystems cater to tribalism (O'Reilly, 2018). In such an environment, dehumanization thrives, making violence—both rhetorical and physical—seem not only possible but justifiable. David Livingstone Smith, in his pivotal work *Making Monsters: The Uncanny Power of Dehumanization*, underscores this dynamic, pointing out how these mechanisms allow people to suppress their natural empathic responses, justifying brutality with ease (Smith, 2021).

But dehumanization also has a darker, more psychological undercurrent. It relates to what Julia Kristeva calls *the abject*—the disturbing, ambiguous border that separates the self from the Other. Kristeva, in her seminal work *Powers of Horror*, defines the abject as that which is cast out, rejected, yet still haunts us (Kristeva, 2024). It is the *not-quite-human*, the figure that both repulses and fascinates because it blurs the boundary between the acceptable and the unacceptable, the familiar and the alien. The abject is what society must expel to maintain its sense of order and identity, yet

 DOI: 10.1201/9781003624813-12

it is never fully eradicated. It lingers on the edges, threatening to return and disrupt the fragile coherence of the self.

In this sense, dehumanization is not just about power or exclusion; it is also about confronting the abject—those aspects of humanity that we cannot fully accept but cannot completely eliminate. Whether it's the image of refugees huddled in make-shift camps, the bodies of victims of war and poverty, or even the stigmatized groups within our own societies, these figures represent the abject. They are reminders of our shared vulnerability, mortality, and the precariousness of the boundaries that define us as 'human' (Kristeva, 2024).

Kristeva's abject can be seen in how societies react to those on the margins. The refugee, the homeless, the racialized "Other," all provoke a visceral reaction that is not just about fear of the unfamiliar but fear of contamination—of the breakdown of the categories we cling to for our sense of identity. When we dehumanize these groups, we are not merely denying their humanity; we are also expelling the parts of ourselves that we find most unsettling.

This theme of exclusion and abjection runs throughout history. Colonialism, slavery, genocide—these were not just acts of domination but also attempts to cast entire groups of people into the realm of the abject, outside the boundaries of humanity. The colonial project, in particular, relied on constructing the colonized as abject—neither fully human nor fully animal, but something in between, something monstrous. This framing allowed empires to justify the most horrific acts of violence, all while maintaining the illusion of moral superiority (Smith, 2021).

The Transatlantic Slave Trade, the Holocaust, and the Rwandan Genocide—each event stands as a horrific reminder of how fragile the concept of humanity can be. In the case of African slavery, European empires systematically dehumanized millions of people to justify their brutal exploitation. This dehumanization wasn't just in the rhetoric of the time; it was embedded in legal systems, economic structures, and cultural practices that allowed entire nations to flourish while denying the humanity of those whose labor and suffering fueled that prosperity.

Even the philosophical movements that promised universal human rights often did so with severe limitations. The Enlightenment, for all its contributions to the idea of human equality, was still deeply entangled with colonialism and the racial hierarchies that accompanied it (Hirschmann, 2002). Figures like Immanuel Kant and Thomas Jefferson, despite their contributions to ideas of freedom and human dignity, held racist views that excluded entire groups from their visions of liberty and equality.

In today's world, dehumanization and abjection persist—perhaps more insidiously than ever—hidden within our digital platforms, economic structures, and political discourse. The abject remains an undercurrent in how we treat migrants, the homeless, or even entire populations caught in conflict zones. Certain lives are still deemed less worthy of dignity, protection, or even basic empathy. We see it in the narratives that surround migrants, portrayed as criminals or invaders, as if their very existence poses a threat to the "humanity" of those within the borders they seek to cross (Zuboff, 2019).

This exclusionary logic extends beyond human beings. Our relationship with the non-human world is also defined by hierarchies of value, as we have systematically dehumanized and commodified animals, ecosystems, and entire species (Eldridge, 2020). Climate change and environmental destruction are, at their core, crises of dehumanization. By placing humans—and, more specifically, wealthy, industrialized humans—at the top of a hierarchical order, we have justified the exploitation and destruction of the very systems that sustain life. In this sense, dehumanization is not just about how we treat other people; it's about how we view our place in the world and our relationship with the non-human.

And yet, despite the weight of history and the deeply ingrained systems that perpetuate exclusion, there are moments of hope—instances where humanity has been expanded rather than contracted. The abolition of slavery, the civil rights movement, the fight for LGBTQ+ rights, and the global struggle for environmental justice are all examples of humanity stretching its definitions, challenging the boundaries of who is considered deserving of dignity, rights, and empathy. These movements, though often met with violent resistance, are testaments to the fact that humanity is not a fixed category but a fluid one—one that can be redefined and expanded through collective action.

As we approach The End (of Humanity?), we must recognize that this process is far from over. The commodification of identity, the policing of borders, and the exploitation of the most vulnerable all continue to thrive on dehumanizing logics. While the tools have changed, the function remains the same. This brings us to the uncomfortable reality: humanity has always been a selective term, a status granted only to some and denied to others.

THE END (OF HUMANITY?)

Transhumanism and post-humanism are like two estranged cousins at the family reunion of philosophy, each vying to redefine humanity in the age of technological marvels. Transhumanism dazzles with promises of human enhancement—prolonging life, amplifying intelligence, and boosting physical prowess through technology (Bostrom, 2014). It's the futurist's toolkit for crafting a superior version of ourselves. Meanwhile, post-humanism invites us to zoom out from our narcissistic human-centric lens, nudging us to ponder: What does it really mean to be "human"?

Post-humanism pushes the envelope, asking us to reconsider humanity's identity as technology evolves (Ferrando, 2013). It boldly confronts anthropocentrism, challenging the assumption that humans are the crown jewel of intelligence and capability. With artificial intelligence (AI) outpacing human cognitive abilities in decision-making and creativity, we are no longer the uncontested rulers of the intellectual realm (Bostrom, 2014). This evolution forces us to wrestle with weighty questions: Are we on the brink of a world where humans are no longer the apex of intelligence? Can traditional concepts of agency and morality survive in a landscape where machines rival human thought?

Going through this complex terrain demands more than just technical acumen; it requires a profound ethical discourse on AI's role in shaping our future. Are these machines our allies in enhancing human qualities, or are they quietly usurping them (Gunkel, 2024)? As we embed AI and other technologies into our societal fabric, ethical design and implementation that prioritize the well-being of all life forms become imperative. Engaging diverse perspectives and communities enriches this discourse, ensuring that the future we shape is inclusive and interconnected.

Post-humanism calls for a reimagined worldview that transcends archaic boundaries, advocating for a future that honors the symbiosis of all life—human and non-human alike (Wolfe, 2009). This inclusive narrative challenges us to acknowledge our shared existence and collective responsibilities, nudging us toward a more holistic ethical framework. This perspective isn't just a futuristic daydream; it's deeply rooted in our history. We've never been purely "natural" beings. Our existence has always intertwined with the tools we create—writing, machinery, and digital networks. These aren't mere extensions but vital parts of our being, constantly reshaping what it means to be human (Ihde, 1990). This historical lens helps us grasp the significance of our current technological trajectory and its implications for humanity's future.

In this redefined human landscape, the posthuman emerges not as a rejection of humanity but as its evolutionary leap (Graham, 2002). It invites us to break down the walls that separate us from other beings—whether they be animals, machines, or even barbarians. This isn't just a nostalgic nod to our origins but a revolutionary call to transcend outdated distinctions, cultivating a comprehensive understanding of life's interconnectedness. It's a perspective that propels us to reevaluate our ethical compass and values in an ever-complicated, interdependent world.

The posthuman isn't about discarding our humanity; it's about expanding its horizons. As we confront the rapid advancements of AI and technology, the posthuman compels us to rethink what it means to be human in this swiftly changing world. Embracing this paradigm doesn't diminish our humanity—it amplifies it, guiding us toward a more equitable and sustainable existence. In this newfound understanding, humanity's essence reveals itself not in rigid definitions but in the depth of our empathy, creativity, and collaboration.

Posthumanism also prompts us to revisit our constructs of humanity through a global and historical lens, illuminating humanity's perennial fascination with hybrid beings and the interplay of nature and culture across civilizations (Braidotti, 2019). While often framed as a modern response to technological advancement, the concept of blending human with non-human traits has ancient roots.

Consider ancient Egypt, where the sphinx—part human, part lion—symbolized a complex fusion of the mortal and the divine. In Mesoamerican cultures, hybrid beings like the jaguar-human figures embodied a spiritual and physical unity that blurred the lines between humans and animals. Similarly, African cosmologies often depict deities as hybrids, representing the seamless coexistence of human and non-human realms (Hastings, 1976). These ancient narratives suggest that humanity has always been a fluid concept, constantly reshaped by its relationship with the non-human. The Renaissance, too, indulged in this exploration, with artists like Hieronymus Bosch creating surreal worlds populated by hybrid creatures that reflected humanity's fears and fascinations with the unnatural. These artistic explorations resonate with today's posthumanist discourse, hinting at an ongoing narrative where humanity is perpetually in flux.

Incorporating these cultural legacies into modern posthumanism enriches our understanding of hybridity (Braidotti, 2019). The rise of AI and biotechnology is merely the latest chapter in humanity's long-standing story of transformation. The challenges we face today are the modern echoes of ancient contemplations on the boundaries of human existence. Non-Western traditions offer invaluable insights into this discourse. Many have long viewed humanity as an integral part of a broader web of life, where technology, nature, and the supernatural coexist fluidly (Mignolo, 2007). This perspective is crucial as we navigate the complexities of a posthuman future, providing a richer, more nuanced understanding of our place in the world.

Thus, posthumanism is not a simple leap into the future; it's a deep dive into the fluidity of human identity. As we face technological frontiers, we draw from a legacy that has always seen humanity as adaptable and interconnected. This historical continuity reminds us that the future of humanity is not a departure from the past but a continuation of our ongoing journey of self-discovery and transformation.

In this light, posthumanism isn't just about redefining the future; it's about uncovering the ancient, evolving essence of humanity itself. As we forge ahead, we carry with us the understanding that humanity's boundaries have always been flexible, shaped by our relationship with the non-human world. The posthuman future beckons not as an end but as a new chapter in our timeless story of adaptation and integration.

REWRITING HUMANITY

The relentless march of technological evolution compels us to redefine what it means to be human. For centuries, we've prided ourselves on traits like rationality, emotional depth, creativity, and social connection—qualities that supposedly set us apart from the rest of life's grand parade. But as we stand on the brink of breakthroughs in genetic engineering, AI, and neural enhancements, one can't help but ask: Do these hallmark traits still belong solely to humans?

This brave new world of bioengineered organisms and AI companions isn't just the stuff of sci-fi novels. It's a tectonic shift in the social landscape, where the boundaries between the human and the artificial begin to dissolve (Harari, 2018). Imagine AI-driven companions that mimic empathy or bioengineered beings with sharper cognitive prowess. These aren't merely curiosities; they're harbingers of a future where our interactions—and even our affections—might extend beyond traditional human connections. Suddenly, the concept of companionship is up for a radical rethink.

With these advancements come weighty ethical conundrums. If an AI can "feel" or a bioengineered entity can "think," what rights do they deserve (Bostrom, 2014)? How do we slot these new beings into our centuries-old moral frameworks? The lines between human and non-human blur, nudging us toward a more inclusive, perhaps even more chaotic, understanding of humanity.

Imagine the societal upheaval when we begin recognizing emotional capacity in machines or sentience in genetically modified organisms. This could upend our sense of belonging and community, pushing us to extend our empathy in directions previously unconsidered (Turkle, 2011). Suddenly, the circle of who—or what—deserves care and companionship gets a whole lot bigger.

As we redefine humanity, our social structures must evolve. The proliferation of diverse identities—whether born from technological tinkering, genetic modification, or entirely new forms of consciousness—demands a fresh culture of empathy and inclusivity. We're rewriting the social contract, advocating for the rights of beings that may look nothing like us but share the ability to experience life in profound ways.

Ultimately, this reimagining of humanity isn't about diluting our essence but enriching it. It's an invitation to transcend historical limitations and widen our moral horizons. As we wade deeper into this era of integration between biology and technology, we must engage in critical, and often uncomfortable, conversations about identity, ethics, and belonging. By doing so, we don't just acknowledge the complex web of existence we're part of—we embrace it. In this shared quest for meaning and connection, humans, machines, and all forms of life become co-authors of a future that's as inclusive as it is uncertain.

WHAT DOES IT MEAN TO BE HUMAN IN 2125?

Imagining the year 2125 invites us into a world where AI and advanced technologies are seamlessly integrated into daily life, reshaping our very understanding of humanity. In this futuristic landscape, people may enhance their cognitive and physical abilities through groundbreaking tech, fundamentally redefining what it means to be human. We could see the rise of hybrid beings—entities that fuse human traits with machine-like efficiency, sparking profound questions about identity and existence.

In this future, AI could become a vital partner in various spheres, significantly transforming healthcare, education, and even our personal relationships. Picture AI systems that continuously monitor our health, providing personalized treatment plans based on our unique genetic makeup and lifestyle choices, potentially leading to longer, healthier lives. In education, intelligent tutoring systems could adapt to each learner's style, ensuring everyone receives tailored educational experiences that maximize their potential (Luckin & Holmes, 2016). Our relationships could also evolve, with AI companions offering emotional support, helping to combat loneliness and foster meaningful connections.

But this promising vision isn't without its shadows. The rapid integration of AI brings critical challenges regarding privacy, autonomy, and the potential erosion of what it means to live a fulfilling life. As we hand over more decision-making power to AI systems, we must confront the implications for personal agency and responsibility. Will we lose touch with our core values and desires, becoming passive consumers of algorithmic recommendations? The danger lies in becoming so reliant on AI that we end up conforming to standardized metrics of success that may not align with our individual identities.

Additionally, the 2125 landscape could be marred by stark disparities in access to these transformative technologies. Society could split into two camps: those who can afford the latest enhancements and those left behind without even basic resources (Brynjolfsson & McAfee, 2014). This growing technological divide could exacerbate existing inequalities, creating a chasm between the "enhanced" and the "unenhanced," challenging our commitment to equity and justice.

Facing this complex future demands a collective effort to ensure technology enhances rather than diminishes the human experience. We must actively engage in ethical discussions about AI, advocating for frameworks that prioritize human dignity, diversity, and inclusion (Gunkel, 2024). Standing on the brink of this new era, we must be vigilant in shaping a future that reflects our highest ideals, ensuring that the advancements we adopt enrich our humanity rather than erode it.

So, what will our shared humanity look like in 2125? Will we rise to the challenge, fostering a society that thrives on collaboration between humans and machines, or will we drift away from our core identities? The choices we make today will shape not just our future interactions with technology but also the very essence of what it means to be human in a rapidly evolving world.

ARE WE ALREADY POSTHUMAN?

The question of whether humanity has ever existed in a purely "human" state challenges us to explore the intricate history of our relationship with technology. From the first primitive tools to the rise of the internet and AI, technology has always played a pivotal role in shaping human identity (Verbeek, 2005). This ongoing interaction suggests a fascinating idea: perhaps we've always existed on a continuum with our technological creations, hinting that we've never truly been "purely" human at all.

Throughout history, every technological breakthrough has not only changed how we interact with the world but also how we relate to each other. Take writing, for example—a revolutionary innovation that transformed communication and knowledge preservation (Ong, 1982). This pivotal moment reshaped entire civilizations, allowing ideas to transcend generations and geographic boundaries. Writing made it possible to document thoughts and experiences, creating a shared history and laying the foundation for collective identity that stretched across time.

Fast-forward to today, where technologies like social media and AI are reshaping our social structures, emotional bonds, and self-perceptions in ways we've never seen before. The rise of social media has blurred the lines between public and private life, fostering a new kind of identity often designed for external approval (Boyd, 2014). In this world, our worth is increasingly measured in likes and shares, making us question the authenticity of our interactions and the very essence of our individuality.

This historical lens forces us to ask whether our identities have ever been purely natural or whether they've always been intertwined with the technologies we've created. Our technological extensions—whether it's writing, machinery, or digital networks—are not just tools; they are core components of our existence that continually redefine what it means to be human (Heidegger, 1977). Acknowledging this long-standing relationship with technology allows us to understand the profound implications of our current trajectory and how it shapes the collective future we're heading into.

LIVING IN A HYBRID WORLD

Posthumanism compels us to critically examine traditional constructs of humanity, suggesting that our understanding of what it means to be human is not static but evolves alongside technological advancements. This philosophical movement challenges the anthropocentric perspective, highlighting how AI, robotics, and biotechnology fundamentally reshape our self-conception. As AI systems increasingly surpass human capabilities in cognitive tasks, decision-making, and even creativity, we must confront the profound implications of these developments for our identity and agency.

Entering an era defined by the coexistence of humans and intelligent machines requires us to embrace this hybrid reality with a focus on ethical considerations and inclusivity. This unique dynamic presents both challenges and opportunities, necessitating a reevaluation of our values and principles and prompting us to reconsider the very essence of our humanity.

Central to this discourse is the concept of *Homo Constructus*, which I propose as a redefinition of what it means to be human in this hybrid world. Rather than adhering to rigid classifications based solely on self-recognition, Homo Constructus invites us to define humanity through self-questioning and the recognition of others. This framework emphasizes that our identity is shaped not only by how we perceive ourselves but also by how we interact with those around us, including non-human entities—whether biological or artificial.

This perspective encourages a more inclusive and expansive understanding of humanity, where the boundaries between humans, animals, and machines are fluid and permeable. By defining ourselves as beings who recognize and engage with others—regardless of their form—we challenge the traditional human/non-human dichotomy. As Donna Haraway aptly states, "we have never been purely human"; we have always been a hybrid construct, shaped by our relationships with the world around us.

The integration of intelligent machines into our daily lives raises essential questions about identity and agency. How do we navigate the complexities of self in a world where AI assists and interacts with us on emotional and cognitive levels? Acknowledging that AI can enhance our capabilities encourages a collaborative mindset, allowing us to leverage the strengths of both humans and machines. This partnership offers an opportunity to redefine societal values in ways that prioritize empathy, compassion, and social responsibility.

Living in a hybrid world also necessitates a commitment to inclusivity. As technology becomes increasingly pervasive across sectors like healthcare and education, we must ensure that these advancements benefit everyone, regardless of socioeconomic status, culture, or background. A hybrid future should not exacerbate existing inequalities but instead promote a more equitable society where the advantages of technology are accessible to all.

In this evolving technological landscape, it is crucial to engage in ongoing discussions about the ethical implications of our choices. What values should guide our interactions with autonomous systems? As automation reshapes our world, we must foster a culture that prioritizes empathy and social responsibility. These questions lie at the heart of the critical dialogue necessary to navigate the complexities of our hybrid existence.

A key challenge in this discussion is defining what it means to be human. Philosopher Giorgio Agamben asserts that self-recognition is a prerequisite for humanity, complicating this issue further. While this notion of self-recognition may seem inclusive, it often reflects Eurocentric and patriarchal assumptions that historically exclude those outside narrowly defined categories of race, gender, or culture.

Linnaeus's taxonomical system, for instance, framed self-recognition within a Western lens, dehumanizing those who did not conform to European ideals.

Similarly, the term *humanitas* was contrasted with *homo barbarus*, setting apart the cultured citizen from the uncivilized outsider. Thus, the concepts *humanus* and *Homo sapiens* are not neutral, scientific labels; they carry the weight of historical exclusion, embedded not only in their Latin etymology but also in the broader scientific and social hierarchies they helped to perpetuate.

Yet, the concept of *Homo Constructus* offers a pathway to move beyond these exclusionary definitions. By embracing this framework, we acknowledge that humanity is not a static label but an evolving construct shaped by our relationships and interactions. In this light, the challenge lies not only in defining humanity but in reprogramming our understanding to reflect a more inclusive and equitable vision.

So, as we log off, shut down, and press Ctrl+Alt+TheEnd, we confront a final, urgent question: Can we reprogram the system before it's too late? Can we rewrite the code that defines humanity to ensure that no one is left out or erased in the process?

Bibliography

Ades, D. (1974). *Dada and Surrealism Reviewed*. The University of Chicago Press.

Agamben, G. (1998). *Homo Sacer: Sovereign Power and Bare Life*. Stanford University Press.

Agamben, G. (2004). *The Open: Man and Animal*. Stanford University Press.

Al-Ghazali, A. (2002). *The Incoherence of the Philosophers*. Brigham Young University.

Al-Arabi, I. (1980). *The Bezels of Wisdom* (A. J. Arberry, Trans.). Paulist Press.

Algorithmic Justice League. (2021). *Algorithmic Justice League: About*. https://www.ajl.org

Allen, W. (Director). (1973). *Sleeper* [Film]. United Artists.

American Trucking Associations. (2019). *The Impact of Automation on the Trucking Industry*. American Trucking Associations. https://www.trucking.org

Amnesty International. (2019). *Ethical AI Principles Won't Solve a Human Rights Crisis*. Amnesty International.

Amodei, D., Olah, C., Steinhardt, J., Christiano, P., Schulman, J., & Man, D. (2016). Concrete Problems in AI Safety. *ArXiv*. https://arxiv.org/abs/1606.06565

Anderson, B. (1983). *Imagined Communities: Reflections on the Origin and Spread of Nation alism*. Verso.

Angwin, J., Larson, J., Mattu, S., & Kirchner, L. (2016). *Machine Bias: There's Software Used across the Country to Predict Future Criminals. And It's Biased against Blacks*. ProPublica.

Appiah, K. A. (2007). *Cosmopolitanism: Ethics in a World of Strangers*. W. W. Norton & Company.

Aristotle. (1995). *Politics* (E. Barker, Trans.). Oxford University Press.

Aristotle. (2020). *Nicomachean Ethics* (A. Beresford, Trans.). Penguin Classics.

Arute, F., Arya, K., Babbush, R. et al. (2019). Quantum Supremacy Using a Programmable Superconducting Processor. *Nature, 574*(7779), 505–510.

Babbush, R., Gidney, C., Berry, D. W., Wiebe, N., McClean, J., Paler, A., Fowler, A., & Neven, H. (2018). Encoding Electronic Spectra in Quantum Circuits with Linear T Complexity. *Physical Review X, 8*(4), 041015.

Badham, J. (Director). (1983). *Wargames* [Film]. United Artists.

Bainbridge, W. S. (2018). Detroit: Become Human and the Ethics of AI. In *The Oxford Handbook of Ethics of AI* (pp. 145–160). Oxford University Press.

Bakshy, E., Messing, S., & Adamic, L. A. (2015). Exposure to Ideologically Diverse News and Opinion on Facebook. *Science, 348*(6239), 1130–1132. https://doi.org/10.1126/science.aaa1160

Baldé, C. P., et al. (2015). The Global E-Waste Monitor 2014. *United Nations University*. https://i.unu.edu/media/unu.edu/news/52624/UNU-1stGlobal-E-Waste-Monitor-2014-small.pdf

Bambach, C. (2003). *Leonardo da Vinci and the Anatomy of the Human Form*. Yale University Press.

Barratt, T., Goods, C., & Veen, A. (2023). Australia: Labour and the Gig Economy. In I. Ness (Ed.), *The Routledge Handbook of the Gig Economy* (1st ed., pp. 347–358), Taylor & Francis.

Barthes, R. (1975). *The Pleasure of the Text*. (R. Miller, Trans.). Hill and Wang.

Barthes, R. (1977). *Image-Music-Text*. Fontana Press.

Bayne, T., Hohwy, J., & Owen, A. M. (2020). *The Oxford Handbook of Consciousness*. Oxford University Press.

Bellwood, P. (2005). *First Farmers: The Origins of Agricultural Societies*. Blackwell Publishing.

Bennett, C. J., & Raab, C. D. (2006). *The Governance of Privacy and Data Protection in the Digital Age*. MIT Press.

Bennett, T. (2015). *Cuts and Criminality: Body Alteration in Legal Discourse*. Routledge.

Benson, C. D. (1982). *The History of the "Roman de la Rose"*. University of Pennsylvania Press.

Berkes, F. (2018). *Sacred Ecology*. Routledge.

Bertot, J. C., Jaeger, P. T., & Grimes, J. M. (2010). Using ICTs to Create a Culture of Transparency: E-Government and Social Media as Openness and Anti-Corruption Tools for Societies. *Government Information Quarterly, 27*(3), 264–271. https://doi.org/10.1016/j.giq.2010.03.001

Bhambra, G. K. (2014). *Connected Sociologies*. Bloomsbury.

Binns, R. (2018a). Algorithmic Accountability and Public Reason. *Philosophy & Technology, 31*(4), 543–556. https://doi.org/10.1007/s13347-017-0263-5

Binns, R. (2018b). La justicia en el aprendizaje automático: Lecciones de filosofía política. In *Actas de la 1ª Conferencia sobre Equidad, Responsabilidad y Transparencia*, vol. 81, 149–159. PMLR.

Biondi, M., & Phillips, E. (2019). The Evolution of Prosthetic Technology: Bridging the Gap between Human and Machine. *Journal of Medical Ethics, 45*(5), 299–305.

Black, E. (2008). *IBM and the Holocaust: The Strategic Alliance between Nazi Germany and America's Most Powerful Corporation*. Dialog Press.

Blaise, N. (Director). (2013). *Elysium* [Film]. Tristar Pictures.

Boccioni, U. (1913). *Unique Forms of Continuity in Space* [Bronze sculpture]. Museo del Novecento, Milan, Italy.

Boden, M. A. (1998). Creativity and Artificial Intelligence. *Artificial Intelligence, 103*(1–2), 347–356. https://philpapers.org/rec/BODCAA-6

Bonsay, J., Cruz, A. P., Firozi, H. C., & Camaro, P. J. C. (2021). Artificial Intelligence and Labor Productivity Paradox: The Economic Impact of AI in China, India, Japan, and Singapore. *Journal of Economics Finance and Accounting Studies, 3*(2), 120–139. https://doi.org/10.32996/jefas.2021.3.2.13

Borenstein, J., Herkert, J. R., & Miller, K. W. (2019). Self-Driving Cars and Engineering Ethics: The Need for a System Level Analysis. *Science and Engineering Ethics, 25*(2), 383–98. https://doi.org/10.1007/s11948-017-0006-0

Boresta, M., Croella, A. L., Gentile, C., Palagi, L., Pinto, D. M., Stecca, G., & Ventura, P. (2024). Optimal Network Design for Municipal Waste Management: Application to the Metropolitan City of Rome. *Logistics, 8*(3), 79. https://doi.org/10.3390/logistics8030079

Bostrom, N. (2014). *Superintelligence: Paths, Dangers, Strategies*. Oxford University Press.

Boyce, M. (2001). *Zoroastrians: Their Religious Beliefs and Practices*. Routledge.

Boyd, D. (2014). *It's Complicated: The Social Lives of Networked Teens*. Yale University Press.

Braidotti, R. (2013). *The Posthuman*. Polity Press.

Braidotti, R. (2019). *Posthuman Knowledge*. Columbia University Press.

Braudel, F. (1972). *The Mediterranean and the Mediterranean World in the Age of Philip II*. Harper & Row.

Brey, P., & Dainow, B. (2023). Ethics by Design for Artificial Intelligence. *AI and Ethics, 4*(4), 1265–1277.

Briggs, R., & King, T. J. (1952). Transplantation of Living Nuclei from Blastula Cells into Enucleated Frogs' Eggs. *Proceedings of the National Academy of Sciences of the United States of America, 38*(5), 455–463.

Broughton, M., Verdon, G., McCourt, T., Martinez, Yoo, J. H., Isakov, S. V., Massey, P., Halavati, R., Niu, M. Y., Zlokapa, A., Peters, E., Lockwood, O., Skolik, A., Jerbi, S., Dunjko, V., Leib, M., Streif, M., Von Dollen, D., Chen, H., Cao, S., Wiersema, R., Huang, H.-Y., McClean, J. R., Babbush, R., Boixo, S. Bacon, D., Ho, A. K., Neven, H., & Mohseni, M. (2021). TensorFlow Quantum: A Software Framework for Quantum Machine Learning. *arXiv*. https://arxiv.org/abs/2003.02989

Brown, D. (2019). *Anubis: The Jackal-Headed God of Egypt*. Oxford University Press.

Brynjolfsson, E., & McAfee, A. (2014). *The Second Machine Age: Work, Progress, and Prosperity in a Time of Brilliant Technologies*. W. W. Norton & Company.

Bryson, J. J. (2010). Robots Should Be Slaves. In Y. Wilks (Ed.), *Close Engagements with Artificial Companions* (pp. 63–74). John Benjamins.

Budge, E. A. W. (1967). *Egyptian Religion: Egyptian Ideas of the Future Life*. Dover Publications.

Bukatman, S. (1997). *Blade Runner: Cyberpunk, Postmodernism, and the Filmic Imagination*. Routledge.

Burge, T. (2010). *Origins of Objectivity*. Oxford University Press.

Burrell, J. (2015). How the Machine 'Thinks:' Understanding Opacity in Machine Learning Algorithms. (September 15, 2015). http://dx.doi.org/10.2139/ssrn.2660674

Cadwalladr, C., & Graham-Harrison, E. (2018). The Cambridge Analytica Scandal and the Exploitation of Personal Data. *The Guardian*. Retrieved from https://www.theguardian.com

Calo, R. (2015). Robotics and the Lessons of Cyberlaw. *California Law Review, 103*(3), 513–563.

Campbell, J. (2008). *The Hero with a Thousand Faces*. Princeton University Press.

Cao, Y., Jonathan Romero, Jonathan P. Olson, Matthias Degroote, Peter D. Johnson, Mária Kieferoválan, D. Kivlichan, Tim Menke Borja Peropadre Nicolas P. D. Sawaya, Sukin Sim, Libor Veis & Alán Aspuru-Guzik. (2019). Quantum Chemistry in the Age of Quantum Computing. *Chemical Reviews, 119*(19), 10856–10915. https://doi.org/10.1021/acs.chemrev.8b00803

Carr, N. (2010). *The Shallows: What the Internet Is Doing to Our Brains*. W. W. Norton & Company.

Carr, N. (2014). *The Glass Cage: How Our Computers Are Changing Us*. W. W. Norton & Company.

Cascio, W. F., & Montealegre, R. (2016). How Technology is Changing Work and Organizations. *Annual Review of Organizational Psychology and Organizational Behavior, 3*, 279–301.

Castells, M. (1996). *The Rise of the Network Society*. Blackwell Publishers.

Cath, C. (2018). Governing Artificial Intelligence: Ethical, Legal and Technical Opportunities and Challenges. *Philosophical Transactions of the Royal Society A: Mathematical, Physical and Engineering Sciences, 376*(2133), 20180080. https://doi.org/10.1098/rsta.2018.0080

Cerezo, M., Andrew Arrasmith, Ryan Babbush, Simon C. Benjamin, Suguru Endo, Keisuke Fujii, Jarrod R. McClean, Kosuke Mitarai, Xiao Yuan, Lukasz Cincio & Patrick J. Coles. (2021). Variational Quantum Algorithms. *Nature Reviews Physics, 3*, 625–644. https://doi.org/10.1038/s42254-021-00348-9

Chalmers, D. J. (1996). *The Conscious Mind: In Search of a Fundamental Theory*. Oxford University Press.

Chalmers, D. J. (2010). *The Character of Consciousness*. Oxford University Press.

Chalmers, D. J. (2016). Panpsychism and the Hard Problem of Consciousness. *Journal of Consciousness Studies, 23*(1), 64–70.

Childe, V. G. (1951). *Man Makes Himself*. New American Library.

Chingono, N. (2024, September 17). Zimbabwe to Cull 200 Elephants to Feed People Left Hungry by Drought. Reuters. https://www.reuters.com/world/africa/zimbabwe-cull-200-elephants-feed-people-left-hungry-by-drought-2024-09-17/

Chomsky, N. (1986). *Knowledge of Language: Its Nature, Origin, and Use*. Praeger.

Chomsky, N. (2006). *On Language and Mind*. Cambridge University Press.

Chomsky, N. (2013). *The Continuity of Language and the Nature of the Human Mind*. Harvard University Press.

Chomsky, N. (2017). *What Kind of Creatures Are We?* Columbia University Press.

Chomsky, N. (2021). *The Role of Language in Human Cognition*. MIT Press.

Chomsky, N. (2023). The False Promise of ChatGPT. *The New York Times*. https://www.nytimes.com/2023/03/08/opinion/noam-chomsky-chatgpt-ai.html

Chui, J., Manyika, J., & Miremadi, M. (2016). Where Machines Could Replace Humans—And Where They Can't (Yet). *McKinsey Quarterly*. https://www.mckinsey.com/capabilities/mckinsey-digital/our-insights/where-machines-could-replace-humans-and-where-they-cant-yet

Cicero. M. T. (1991). *On Duties* (M. T. Griffin & E. M. Atkins, Eds. & Trans.). Cambridge University Press.

Cicero. M. T. (2005). *On the Republic* (M. T. Granger, Ed.). Cambridge: Cambridge University Press.

Clark, A. (2003). *Natural-Born Cyborgs: Minds, Technologies, and the Future of Human Intelligence*. Oxford University Press.

Clarke, S., Savulescu, J., Coady, T., Giubilini, A., & Sanyal, S. (Eds.). (2016). *The Ethics of Human Enhancement: Understanding the Debate*. Oxford University Press.

Clynes, M. E., & Kline, N. S. (1960). Cyborgs and Space. *Astronautics*, 14(9), 26–27, 74–76.

Cohen, E. (2009). *A Body Worth Defending: Immunity, Biopolitics, and the Apotheosis of the Modern Body*. Duke University Press.

Colton, S. (2012). The Painting Fool: Stories from Building an Automated Painter. In J. McCormack & M. d'Inverno (Eds.), *Computers and Creativity* (pp. 3–38). Springer.

Confucius. (2003). *The Analects* (E. Slingerland, Trans.). Hackett Publishing.

Content CES. (2021). *AI-Enabled Robotic Weeders in Precision Agriculture*. North Carolina State University. https://content.ces.ncsu.edu/Artificial-Intelligence-Ai-Enabled-Robotic-Weeders-In-Precision-Agriculture

Cormen, T. H., Leiserson, C. E., Rivest, R. L., & Stein, C. (2009). *Introduction to Algorithms* (3rd ed.). MIT Press.

Crawford, K. (2021). *Atlas of AI: Power, Politics, and the Planetary Costs of Artificial Intelligence*. Yale University Press.

Creanza, N., Kolodny, O., & Feldman, M. W. (2017). Cultural Evolutionary Theory: How Culture Evolves and Why It Matters. *Proceedings of the National Academy of Sciences*, 114(30), 7803–7810.

Daley, G. Q., Scadden, D. T., & Lensch, M. W. (2010). Prospects for Stem Cell-Based Therapy. *Cell, 132*(4), 544–548. https://doi.org/10.1016/J.Cell.2008.02.009

Dalí, S. (1931). *The Persistence of Memory* [Painting]. Museum of Modern Art, New York, NY, United States.

Damasio, A. R. (1999). *The Feeling of What Happens*: Body and Emotion in the Making of Consciousness. Harcourt Brace.

Davis, L. J. (2015). *Enforcing Normalcy: Disability, Deafness, and the Body*. Verso Books.

Dawkins, R. (2006). *The God Delusion*. Bantam Press.

De Waal, F. (2016). *Are We Smart Enough to Know How Smart Animals Are?* W.W. Norton & Company.

Deleuze, G., & Guattari, F. (1987). *A Thousand Plateaus: Capitalism and Schizophrenia* (B. Massumi, Trans.). University of Minnesota Press.

Deloria, V. (2003). *God Is Red: A Native View of Religion*. Fulcrum Publishing.

Derrida, J. (2002). *The Animal That Therefore I Am*. Fordham University Press.

Descartes, R. (1641). *Meditations on First Philosophy* (J. Cottingham, Trans., 1996). Cambridge University Press.

Descartes, R. (1985). *The Philosophical Writings of Descartes* (Vol. 1). Cambridge University Press.

Deutsch, D. (1985). Quantum Theory, the Church-Turing Principle and the Universal Quantum Computer. *Proceedings of the Royal Society A, 400*(1818), 97–117.

Diakopoulos, N. (2016). Accountability in Algorithmic Decision Making. *Data & Society Research Institute*, 59(2), 1–6.

Diamond, J. (1999). *Guns, Germs, and Steel: The Fates of Human Societies*. W.W. Norton & Company.

Dick, P. K. (1968). *Do Androids Dream of Electric Sheep?* Doubleday.

Dickerman, L. (2005). *Dada's Women: The Politics of Gender in the Early Dada Movement*. The Dada Companion.

Digital Government Blueprint. (2018). *Singapore Smart Nation and Digital Government Office*. Smart Nation.

Dissanayake, E. (1992). *Homo Aestheticus: Where Art Comes from and Why*. Free Press.

Doudna, J. A., & Charpentier, E. (2014). The New Frontier of CRISPR Gene Editing. *Science, 345*(6193), 1258424. doi: 10.1126/science.1258096

Dreyfus, H. L. (1992). *What Computers Still Can't Do: A Critique of Artificial Reason*. MIT Press.

Duffy, B. E., & Hund, E. (2015). 'Having It All' on Social Media: Entrepreneurial Femininity and Self-Branding among Fashion Bloggers. *Social Media + Society, 1*(2), 1–11.

Eagleton, T. (2005). *The Function of Criticism*. Verso.

Eberhard, W. (1974). *A Dictionary of Chinese Symbols: Hidden Symbols in Chinese Life and Thought*. Routledge & Kegan Paul.

Eck, D. L. (1998). *Darśan: Seeing the Divine Image in India*. Columbia University Press.

Eisenstein, E. L. (1980). *The Printing Press as an Agent of Change: Communications and Cultural Transformations in Early-Modern Europe*. Cambridge University Press.

Eldridge, P. (2020). *Reconceptualizing Environmental Ethics: Nature, Nonhuman Animals, and the Politics of Life*. Routledge.

Elgammal, A., Liu, B., Elhoseiny, M., & Mazzone, M. (2017). CAN: Creative Adversarial Networks, Generating "Art" by Learning about Styles and Deviating from Style Norms. *Arxiv Preprint Arxiv:1706.07068*.

Eliade, M. (1987). *The Sacred and the Profane: The Nature of Religion*. Harcourt Brace Jovanovich.

Eliot, T. S. (2020). *The Sacred Wood: Essays on Poetry and Criticism*. Cosimo Classics.

Erasmus. (1994). *In Praise of Folly* (B. Radice, Trans.; A. Levi, Intro.). Penguin Classics.

Eubanks, V. (2018). *Automating Inequality: How High-Tech Tools Profile, Police, and Punish the Poor*. St. Martin's Press.

European Commission. (2021). *Proposal for a Regulation of The European Parliament and of the Council Laying Down Harmonized Rules on Artificial Intelligence (Artificial Intelligence Act)*. Brussels.

European Trade Union Confederation (ETUC). (2019). *The Impact of Artificial Intelligence on Labor: A Call for just Transition Policies*. European Trade Union Confederation. Retrieved from https://www.etuc.org

Farah, M. J. (2005). Neuroethics: The Practical and the Philosophical. *Trends in Cognitive Sciences, 9*(1), 34–40.

Faulkner, R. O. (1994). *The Egyptian Book of the Dead: The Book of Going Forth by Day*. Chronicle Books.

Feldman, R. A., Gibbons, J. R., & Miller, W. (2021). Cloning the Black-Footed Ferret: The Future of Conservation Genomics. *Trends in Biotechnology, 39*(5), 431–434.

Ferrando, F. (2013). Posthumanism, Transhumanism, Antihumanism, Metahumanism, and New Materialisms: Differences and Relations. *Existenz, 8*(2), 26–33.

Finocchiaro, M. (2018). *Galileo's Mechanics: A Historical and Philosophical Perspective*. Cambridge University Press.

Flores Mosri, D. (2021). Clinical Applications of Neuropsychoanalysis: Hypotheses Toward an Integrative Model. *Frontiers in Psychology, 12*, 718372. https://doi.org/10.3389/Fpsyg.2021.718372

Flores Mosri, D., Iftah Biran, Richard Kessler and David Olds. (2021). Revisiting Metapsychology, Psychopathology, and Developmental Issues in Neuropsychoanalysis. *Neuropsychoanalysis, 23*(2), 139–157.

Flores Mosri, D., J. Abrams, Virginia C Barry, I. Biran, R. Coetzer, P. Moore, José Fernando Muñoz Zúñiga and Maggie Zellner. (2022). Clinical Writing in Neuropsychoanalysis. *Neuropsychoanalysis, 24*(2), 171–191.

Floridi, L. (2014). *The Fourth Revolution: How the Infosphere is Reshaping Human Reality.* Oxford University Press.

Foucault, M. (1988). *Technologies of the Self: A Seminar with Michel Foucault* (L. Hutton, P. Tarr, & H. Gordon, Eds.). University Of Massachusetts Press.

Foucault, M. (1994). *The Order of Things: An Archaeology of Human Sciences.* Vintage.

Frankl, V. E. (2006). *Man's Search for Meaning* (Rev. ed.). Beacon Press.

Freud, S. (1923). *The Ego and the Id.* SE, *19*, 12–66.

Freud, S. (2003). The Uncanny. Penguin Classics.

Frey, C. B., & Osborne, M. A. (2017). The Future of Employment: How Susceptible are Jobs to Computerisation? *Technological Forecasting and Social Change,* 114, 254–280.

Friedman, J. (1994). *Cultural Identity and Global Process.* Sage Publications.

Fukuyama, F. (2002). *Our Posthuman Future: Consequences of The Biotechnology Revolution.* Farrar, Straus and Giroux.

Fukuyama, F. (2011). *The Origins of Political Order: From Prehuman Times to the French Revolution.* Farrar, Straus and Giroux.

Gadamer, H.- G. (1975). *Truth and Method.* Seabury Press.

Gailhofer, P., Herold, A., Schemmel, J. P., Scherf, C. S., Urrutia, C., Köhler, A. R., & Braungardt, S. (2021). *The Role of Artificial Intelligence in the European Green Deal. Policy Department for Economic, Scientific and Quality of Life Policies, Directorate-General for Internal Policies.* European Parliament. https://data.europa. eu/doi/10.2861/882830

Gamble, C. (1994). *Timewalkers: The Prehistory of Global Colonization.* Harvard University Press.

Garland, A. (Director). (2014). *Ex Machina* [Film]. Universal Pictures.

Geertz, C. (1973). *The Interpretation of Cultures.* Basic Books.

Gibson, W. (1984). *Neuromancer.* Ace Books.

Gilpin, L. H., Bau, D., Yuan, B. Z., Bajwa, A., Specter, M., & Kagal, L. (2018). Explaining Explanations: An Overview of Interpretability of Machine Learning. In *2018 IEEE 5th International Conference on Data Science and Advanced Analytics (DSA*A), pp. 80–89. Turin, Italy. doi: 10.1109/DSAA.2018.00018

Gimbutas, M. (2007). *The Goddesses and Gods of Old Europe: Myths and Cult Images.* University of California Press.

Ginsburg, F., & Rapp, R. (2013). Disability Worlds. *Annual Review of Anthropology,* 42, 53–68.

Giorgi, G. (2020). *The Animal's Turn: Thinking with Animals in the Anthropocene.* University Of Chicago Press.

Gisin, N., Grégoire Ribordy, Wolfgang Tittel, and Hugo Zbinden. (2002). Quantum Crypto graphy. *Reviews of Modern Physics, 74*(1), 145.

Gleick, J. (2011). *The Information: A History, a Theory, a Flood.* Pantheon Books.

Gogoll, J., & Müller, J. (2017). Autonomous Cars: In Favor of a Mandatory Ethics Setting. *Science and Engineering Ethics,* 23(3), 681–700.

Gonlin, N., & Reed, D. M. (Eds.). (2021). *Night and Darkness in Ancient Mesoamerica.* University Press of Colorado.

Goodall, N. J. (2014). Machine Ethics and Automated Vehicles. In *Road Vehicle Automation* (pp. 93–102). Springer Vieweg.

Goodfellow, I., Bengio, Y., & Courville, A. (2016). *Deep Learning.* MIT Press.

Goodman, B., & Flaxman, S. (2018). European Union Regulations on Algorithmic Decision-Making and a "Right to Explanation". *AI & Society,* 33(3), 543–558.

Google. (2020). *AI Principles.* https://ai.google/principles/

Gould, S. J. (1981). *The Mismeasure of Man.* W.W. Norton & Company.

Graham, A. C. (1989). *Disputers of the Tao: Philosophical Argument in Ancient China*. Open Court Publishing.

Graham, E. L. (2002). *Representations of the Post/human: Monsters, Aliens, and Others in Popular Culture*. Rutgers University Press.

Green, Richard E., Johannes Krause, Adrian W. Briggs, Tomislav Maricic, Udo Stenzel, Martin Kircher, Nick Patterson et al. (2010). A draft sequence of the Neandertal genome. *Science* 328(5979), 710–722.

Greenleaf, G. (2023). Global Data Privacy Laws 2023: 162 National Laws and 20 Bills. *181 Privacy Laws and Business International Report (PLBIR)*, 1, 2–4. UNSW Law Research Paper No. 23–48.

Groening, M. (Creator). (1999). *Futurama* [TV Series]. 20th Century Fox Television.

Grover, L. K. (1996). A Fast Quantum Mechanical Algorithm for Database Search. *Proceedings Of The 28th Annual ACM Symposium on Theory of Computing*, pp. 212–219.

Gunkel, D. J. (Ed.). (2024). *Handbook on the Ethics of Artificial Intelligence*. Edward Elgar Publishing.

Habermas, J. (1989) *The Structural Transformation of the Public Sphere: An Inquiry into a Category of Bourgeois Society* (Trans. by Burger T. with the Assistance of Lawrence F.). Polity Press.

Hadot, P. (1995). *The Inner Citadel: The Meditations of Marcus Aurelius*. Harvard University Press.

Halberstam, J. (2011). *The Queer Art of Failure*. Duke University Press.

Hanson Robotics. (2016). *Sophia: The Humanoid Robot That Mimics Human Expressions and Interactions*. https://www.hansonrobotics.com

Hanson Robotics. (2017). *Sophia The Robot: A Conversation with Andrew Sorkin at The Future Investment Initiative*. https://Www.hansonrobotics.com

Harari, Y. N. (2017). *Homo Deus: A Brief History of Tomorrow*. Harper.

Harari, Y. N. (2018). *21 Lessons for the 21st Century*. Spiegel & Grau.

Harari, Y. N. (2024). *Nexus: A Brief History of Information Networks from the Stone Age to AI*. Random House.

Harari, Y. N. (2015). *Sapiens: A Brief History of Humankind*. Harper.

Haraway, D. J. (1991). *Simians, Cyborgs, and Women: The Reinvention of Nature*. Routledge.

Haraway, D. J, (2008). *When Species Meet*. University of Minnesota Press.

Harris, J. (2004a). *On Cloning*. Routledge.

Harris, J. (2004b). *The Ethical Use of Human Embryonic Stem Cells in Research and Therapy*. In A. Dyson & J. Harris (Eds.), *Ethics and Biotechnology* (pp. 165–176). Routledge. https://doi.org/10.1002/9780470756423.Ch12

Harvey, P. (2013). *An Introduction to Buddhism: Teachings, History and Practices*. Cambridge University Press.

Hastings, A. (1976). *African Christianity: An Essay in Interpretation*. G. Chapman.

Hawking, S. (2016). *Brief Answers to the Big Questions*. Bantam Books.

Hayles, N. K. (1999). *How We Became Posthuman: Virtual Bodies in Cybernetics, Literature, and Informatics*. University of Chicago Press.

Heath, R. (2023). *The Water Footprint of AI: A Case Study of ChatGPT*. The Washington Post.

Heidegger, M. (1962). *Being and Time* (J. Macquarrie & E. Robinson, Trans.). Harper & Row.

Heidegger, M. (1977). *The Question Concerning Technology and Other Essays*. Harper & Row.

Heschel, A. J. (1976). *God in Search of Man: A Philosophy of Judaism*. Farrar, Straus and Giroux.

Hesiod. (1999). *Theogony and Works and Days* (M. L. West, Trans.). Oxford University Press.

Hirschmann, A. O. (2002). *The Passions and the Interests: Political Arguments for Capitalism before Its Triumph*. Princeton University Press.

Hobsbawm, E. (1987). *The Age of Revolution: Europe* 1789–1848. Vintage Books.

Hobsbawm, E. (1995). *The Age of Extremes: The Short Twentieth Century, 1914–1991*. Abacus.

Hochberg, L. R., Daniel Bacher, Beata Jarosiewicz, Nicolas Y. Masse, John D. Simeral, Joern Vogel, Sami Haddadin, Jie Liu, Sydney S. Cash, Patrick van der Smagt & John P. Donoghue. (2012). Reach and Grasp by People with Tetraplegia Using a Brain–Machine Interface. *Nature, 485*(7398), 372–375.

Hodder, I. (2011). *Symbols in Action: Ethnoarchaeological Studies of Material Culture.* Cambridge University Press.

Hodder, I. (2012). *Entangled: An Archaeology of the Relationships between Humans and Things.* Wiley-Blackwell.

Huang, G., Ya Wang, Yoo-Geun Ham, Bin Mu, Weichen Tao & Chaoyang Xie. (2024). Toward a Learnable Climate Model in the Artificial Intelligence Era. *Advances in Atmospheric Sciences, 41,* 1281–1288. https://doi.org/10.1007/S00376-024-3305-9

Hughes, T. P. (2004). *Human-Built World: How to Think about Technology and Culture.* University of Chicago Press.

Huidobro, V. (1932). *Altazor: El Viajero Solitario.* Ediciones de la Universidad de Chile.

Huriye, A. Z. (2023). The Ethics of Artificial Intelligence: Examining the Ethical Considerations Surrounding the Development and Use of AI. *American Journal of Technology, 2*(1), 37–44.

Hutton, R. (2001). *The Triumph of the Moon: A History of Modern Pagan Witchcraft.* Oxford University Press.

Huxley, A. (2019). *Brave New World Revisited.* Harper & Row.

IEEE. (2019). *The Algorithm Transparency Standard: Creating Transparency in AI Algorithms.* IEEE Standards Association.

Ihde, D. (1990). *Technology and the Lifeworld: From Garden to Earth.* Indiana University Press.

Imai, M. (1986). *Kaizen: The Key to Japan's Competitive Success.* Random House.

International Labour Organization (ILO). (2021). *The Future of Work in the Age of Automation.* International Labour Organization. https://www.ilo.org

Ishiguro, K. (2006). *Never Let Me Go* (1st ed.). Vintage.

Johanson, D. C., & Edey, M. A. (1981). *Lucy: The Beginnings of Humankind.* Simon & Schuster.

Jung, C. G. (1964). *Man and His Symbols.* Doubleday.

Jurafsky, D., & Martin, J. H. (2021). *Speech and Language Processing: An Introduction to Natural Language Processing, Computational Linguistics, and Speech Recognition* (3rd ed.). Pearson.

Kaes, A. (2010). *Metropolis.* BFI Film Classics.

Kalvet, T. (2012). Innovation: A Factor Explaining E-Government Success in Estonia. *Electronic Government, An International Journal, 9*(2), 142–157.

Kangas, O., Jauhiainen, S., Simanainen, M., & Ylikännö, M. (Eds.). (2021). *Experimenting with Unconditional Basic Income: Lessons from the Finnish BI Experiment 2017–2018.* Edward Elgar Publishing.

Kania, E. B. (2020). Minds at War: China's Pursuit of Military Advantage through Cognitive Science and Biotechnology. *Prism: A Journal of the Center for Complex Operations, 8*(3), 82–101.

Kania, E. B., & Laskai, L. (2021). *Myths and Realities of China's Military-Civil Fusion Strategy.* Center for a New American Security. https://www.cnas.org/Publications/Reports/Myths-And-Realities-Of-Chinas-Military-Civil-Fusion-Strategy

Kant, I. (1781). *Critique of Pure Reason.* Translated by Norman Kemp Smith, Macmillan, 1929.

Kaplan, M. (1990). *The Handbook of Jewish Thought.* Moznaim Pub Corp.

Keenan, J. F. (2009). *Catholic Ethics in Today's World.* Sheed & Ward.

Kemp, M. (2011). *Leonardo da Vinci: The Marvellous Works of Nature and Man.* Harvard University Press.

Kenny, A. (1992). *Aristotle on the Perfect Life.* Clarendon Press.

Kenny, A. (2005). *Aquinas on Being*. Clarendon Press.

Kirkland, R. (2004). *Taoism: The Enduring Tradition*. Routledge.

Kitchin, R. (2016). Thinking Critically about and Researching Algorithms. *Information, Communication & Society, 20*(1), 14–29. https://doi.org/10.1080/1369118X.2016.1154087

Klein, R. G. (2009). *The Human Career: Human Biological and Cultural Origins* (3rd ed.). University of Chicago Press.

Klein, S. (2013). *Bodies of Meaning: Studies on Language, Labor, and Liberation*. State University of New York Press.

Knuth, D. E. (1997). *The Art of Computer Programming* (3rd ed., Vol. 1). Addison-Wesley.

Koch, C. (2018). *The Feeling of Life Itself: Why Consciousness is still a Mystery*. MIT Press.

Kohn, L. (2009). Introducing Daoism. Routledge.

Krauss, R. (1999). *A Voyage on the Sea of the Machine: Art and Technology in the 20th Century*. MIT Press.

Kristeva, J. (2024). *Powers of Horror: An Essay on Abjection*. Columbia University Press.

Kumar, G., Yadav, S., Mukherjee, A., Hassija, V., & Guizani, M. (2024). *Recent Advances in Quantum Computing for Drug Discovery and Development*. IEEE.

Kurzweil, R. (2005). *The Singularity is Near: When Humans Transcend Biology*. Viking.

Kushner, H. (2004). *When Bad Things Happen to Good People*. Anchor Books.

Ladd, A. (1999). *Fritz Lang's Metropolis: Critical Approaches*. Cambridge University Press.

Lang, F. (Director). (1927). *Metropolis* [Film]. Universum Film AG.

Lanza, R. P., Cibelli, J. B., & West, M. D. (1999). Human Therapeutic Cloning. *Nature Medicine, 5*(9), 975–977. doi: 10.1038/12404

Latour, B. (1993). *We Have Never Been Modern*. Harvard University Press.

Latour, B. (2005). *Reassembling the Social: An Introduction to Actor-Network Theory*. Oxford University Press.

Latour, B. (2017). *Facing Gaia: Eight Lectures on the New Climatic Regime*. Polity Press.

Lau, D. C. (2014). *The Analects of Confucius*. Penguin Classics.

Lear, J. (1988). *Aristotle: The Desire to Understand*. Cambridge University Press.

Lebedev, M. A., & Nicolelis, M. A. L. (2006). Brain-Machine Interfaces: Past, Present and Future. *Trends in Neurosciences, 29*(9), 536–546. https://doi.org/10.1016/J.Tins.2006.07.004

Lebreton, L. C. M., Slat, B., Ferrari, F., Sainte-Rose, B., Aitken, J., Marthouse, R., Hajbane, S,. Cunsolo, S., Schwarz, A., Levivier, A., Noble, K., Debeljak, P., Maral, H., Schoeneich-Argent, R., Brambini, R., & Reisser, J. (2018). Evidence that the Great Pacific Garbage Patch is Rapidly Accumulating Plastic. *Scientific Reports, 8*, Article Number: 4666.

Leibniz, G. W. (1998). *Monadology*. Translated by Robert Latta. Open Court Publishing.

Levenson, J. D. (1988). *Creation and the Persistence of Evil: The Jewish Drama of Philosophy*. Princeton University Press.

Lévi-Strauss, C. (1966). *The Savage Mind*. University of Chicago Press.

Levy, D. (2007). *Love and Sex with Robots: The Evolution of Human-Robot Relationships*. HarperCollins.

Lewis, C. S. (1947). *The Abolition of Man*. Macmillan.

Lewis, D. K. (1969). *Convention: A Philosophical Study*. Harvard University Press.

Lindsay, J., & Gartzke, E. (2020). Politics by Many Other Means: The Comparative Strategic Advantages of Operational Domains. *Journal of Strategic Studies*. https://doi.org/10.1080/01402390.2020.1768372

Linnaeus, C. (1758). *Systema Naturae Per Regna Tria Naturae, Secundum Classes, Ordines, Genera, Species, Cum Characteribus, Differentiis, Synonymis, Locis* (10th ed.). Laurentii Salvii.

Liu, X. (2001). The Chinese Dragon: A Symbol of the Divine and Imperial Power. *Asian Folklore Studies, 60*(2), 183–200.

Locatelli, C. (2022). Rethinking 'Sex Robots': Gender, Desire, and Embodiment in Posthuman Sextech. *Gender, Sexuality, and Embodiment in Digital Spheres: Connecting Intersectionality and Digitality, 4*(3), 10–33.

Locke, J. (1689). *Two Treatises of Government.* GreatEbooksCheap.com (24 April 2022).

Longrich, N. (2020). When did we become fully human? What fossils and DNA tell us about the evolution of modern intelligence. The Conversation. https://theconversation.com/when-did-we-become-fully-human-what-fossils-and-dna-tell-us-about-the-evolution-of-modern-intelligence-143717.

Loomba, A. (2015). *Colonialism/Postcolonialism* (3rd ed.). Routledge.

López, J. (2014). *Legal and Ethical Issues in Human Cloning.* Oxford University Press.

Lu, C., Lange, R. T., Foerster, J., Clune, J., & Ha, D. (2024). The AI Scientist: Towards Fully Automated Open-Ended Scientific Discovery. *Computer Science Artificial Intelligence.* https://arXiv:2408.06292v3

Lucas, G. (Director). (1977). *Star Wars: Episode IV – A New Hope* [Film]. 20th Century Fox.

Luckin, R., & Holmes, W. (2016). *Intelligence Unleashed: An Argument for AI in Education.* Pearson Education.

Lum, K., & Isaac, W. (2016). To Predict and Serve? A Look at Predictive Policing. *Significance, 13*(5), 14–19.

Mackenzie, A. (2010). *Wirelessness: Radical Empiricism in Network Cultures.* MIT Press.

Manyika, J., Chui, M., Miremadi, M., Bughin, J., George, K., Willmott, P., & Dewhurst, M. (2017). *A Future That Works: Automation, Employment, and Productivity.* Mckinsey Global Institute.

Marinetti, F. T. (1909). *The Futurist Manifesto. Le Figaro.*

Martin, D. L., Harrod, R. P., & Pérez, V. R. (2013). *Bioarchaeology: An Integrated Approach to Working with Human Remains.* Routledge.

Mawdudi, S. A. (1994). Towards Understanding Islam. Islamic Foundation.

Mayer-Schönberger, V., & Cukier, K. (2013). *Big Data: A Revolution That will Transform How We Live, Work, and Think.* Houghton Mifflin Harcourt.

Mayr, E. (1982). *The Growth of Biological Thought: Diversity, Evolution, and Inheritance.* Harvard University Press.

Mbiti, J. S. (1990). *African Religions and Philosophy* (2nd ed.). Heinemann.

McAdams, D. P. (1993). *The Stories We Live by: Personal Myths and the Making of the Self.* Guilford Press.

McElhenney, R., Howerton, G., & Day, C. (Creators). (2005–present). *It's Always Sunny in Philadelphia* [TV series]. FX.

McGinnis, J. (2005). *Medieval Islamic Philosophical Writings.* Cambridge University Press.

McGrath, A. E. (2011). *Christian Theology: An Introduction* (5th ed.). Wiley-Blackwell.

McInerny, R. (1998). *St. Thomas Aquinas.* University Of Notre Dame Press.

McKee, R. (2023). *The Future of Storytelling: AI and the Writer's Strike.* Creative Writing Press.

Mckibben, B. (2010). *Eaarth: Making a Life on a Tough New Planet.* Times Books.

McLuhan, M. (1964). *Understanding Media: The Extensions of Man.* McGraw-Hill.

Mellars, P. (2006). Why Did Modern Human Populations Disperse from Africa Ca. 60,000 Years Ago? A New Model. *Proceedings of the National Academy of Sciences, 103*(12), 4533–4539.

Microsoft. (2018). *Microsoft's AI Principles: Advancing AI to Empower People.* https://www.microsoft.com/En-Us/Ai/Our-Approach

Microsoft. (2025). *AI For Earth Initiative.* Retrieved From https://www.microsoft.com/En-Us/Sustainability/Emissions-Impact-Dashboard

Mifflin, M. (1997). *Bodies of Subversion: A Secret History of Women and Tattoo.* Juno Books.

Mignolo, W. D. (2007). *The Idea of Latin America.* Blackwell Publishing.

Mignolo, W. D. (2011). *The Darker Side of Western Modernity: Global Futures, Decolonial Options*. Duke University Press.

Miller, B. (2008). *Native America: A Journey into the Heart of the Continent*. Bison Books.

Miller, M., & Taube, K. A. (1993). *The Gods and Symbols of Ancient Mexico and the Maya: An Illustrated Dictionary of Mesoamerican Religion*. Thames & Hudson.

Mithen, S. (1996). *The Prehistory of the Mind: A Search for the Origins of Art, Religion and Science*. Thames and Hudson.

Mithen, S. (2005). *The Singing Neanderthals: The Origins of Music, Language, Mind and Body*. Weidenfeld & Nicolson.

More, M., & Vita-More, N. (Eds.). (2013). *The Transhumanist Reader: Classical and Contemporary Essays on the Science, Technology, and Philosophy of the Human Future* (1st ed., Kindle ed.). Wiley-Blackwell.

Morrison, T. (2017). *Centaurs and the Duality of Identity*. Princeton University Press.

Nagel, T. (1974). What is It Like to Be a Bat? *Journal of Philosophy, 68*(6), 435–450.

Nanda, M. (2003). *Prophets Facing Backward: Postmodern Critiques of Science and Hindu Nationalism in India*. Rutgers University Press.

Nasr, S. H. (1976). *Islamic Science: An Illustrated Study*. Kazi Pubns Inc.

National Institute of Standards and Technology (NIST). (2020). *Post-Quantum Cryptography: Preparing for a Quantum-Safe Future*. NIST.

Nielsen, M. A., & Chuang, I. L. (2010). *Quantum Computation and Quantum Information*. Cambridge University Press.

Nietzsche, F. (1883). Thus Spoke Zarathustra (T. Common, Trans.). Project Gutenberg. https://www.gutenberg.org/files/1998/1998-h/1998-h.htm.

Nolan, J. (Creator). (2016–2022). *Westworld* [TV series]. HBO.

Nussbaum, M. C. (1997). *Cultivating Humanity: A Classical Defense of Reform in Liberal Education*. Harvard University Press.

O'Neil, C. (2016). *Weapons of Math Destruction: How Big Data Increases Inequality and Threatens Democracy*. Crown Publishing Group.

O'Reilly, T. (2018). *WTF?: What's the Future and Why it's Up to Us*. Random House.

Obermeyer, Z., Powers, B. W., Vogeli, C., & Mullainathan, S. (2019). Dissecting Racial Bias in an Algorithm Used to Manage the Health of Populations. *Science, 366*(6464), 447–453. https://doi.org/10.1126/science.aax2342

Omdena. (2022). *Monitoring Reforestation Using Machine Learning*. https://www.omdena.com/blog/monitoring-reforestation-using-machine-learning

Ong, W. (1982). *Orality and Literacy: The Technologizing of the Word*. Methuen & Co. Ltd.

Open AI. (2023). GPT-4 Technical Report. *OpenAI*. https://openai.com/research/gpt-4

Oreskes, N., & Conway, E. M. (2010). *Merchants of Doubt: How a Handful of Scientists Obscured the Truth*. Bloomsbury Press.

Pang, B., & Lee, L. (2008). Opinion Mining and Sentiment Analysis. *Foundations and Trends in Information Retrieval, 2*(1–2), 1–135.

Panksepp, J. (1998). *Affective Neuroscience: The Foundations of Human and Animal Emotions*. Oxford University Press.

Pariser, E. (2011). *The Filter Bubble: What the Internet Is Hiding from You*. Penguin Press.

Park, C. (Director). (2006). *I'm a Cyborg, But That's OK* [Film]. CJ Entertainment.

Partnership on AI. (2020). *Charting a Course Together: 2020 Annual Report*. Partnership on AI.

Perception of AI Art. (2023). *ARISA Foundation*. https://www.arisafoundation.org/perception-of-ai-art

Peruzzo, A., Jarrod McClean, Peter Shadbolt, Man-Hong Yung, Xiao-Qi Zhou, Peter J. Love, Alán Aspuru-Guzik & Jeremy L. O'Brien. (2014). A Variational Eigenvalue Solver on a Quantum Processor. *Nature Communications, 5*, 4213.

Peters, T. (1997). "Goodbye Dolly?" The Ethics of Human Cloning. *Journal of Medical Ethics, 23*(6), 353–360. https://doi.org/10.1136/jme.23.6.353

Picard, R. (2003). Affective Computing: Challenges. *International Journal of Human-Computer Studies,* 59(1–2), 55–64.

Pinker, S. (1997). *How the Mind Works.* W. W. Norton & Company.

Pinker, S. (2018). *Enlightenment Now: The Case for Reason,* Science, Humanism, and Progress. Viking Press.

Pinker, S. (2021). *Rationality: What It is, Why It Seems Scarce,* Why It Matters. Viking Press.

Pitts, J. (2005). *A Turn to Empire: The Rise of Imperial Liberalism in Britain and France.* Princeton University Press.

Plato. (1961). Meno. In *The Collected Dialogues of Plato* (E. Hamilton & H. Cairns, Eds., W. K. C. Guthrie, Trans.). Princeton University Press.

Plautus. (1965). *The Pot of Gold and Other Plays* (E. F. Watling, Trans.). Penguin Classics.

Plautus. (2006). *Asinaria* (W. H. D. Rouse, Trans.). Loeb Classical Library.

Pope John Paul II. (1998). *Evangelium Vitae: The Gospel of Life.* Libreria Editrice Vaticana.

Postman, N. (1993). *Technopoly: The Surrender of Culture to Technology.* Vintage Books.

Pratt, M. L. (1992). *Imperial Eyes: Travel Writing and Transculturation.* Routledge.

Preskill, J. (2018). Quantum Computing in the NISQ Era and Beyond. *Quantum, 2,* 79.

Quijano, A. (2007). Coloniality and Modernity/Rationality. *Cultural Studies, 21*(2–3), 168–178.

Quintilian. (1920). *Institutio Oratoria* (H. E. Butler, Trans.). Loeb Classical Library.

Quirke, S. (1994). *The Death and Afterlife of Ancient Egyptians.* Cambridge University Press.

Rabadán, A. T. (2021). Neurochips: Considerations from a Neurosurgeon's Standpoint. *Surgical Neurology International, 12,* 173.

Rasmussen, M., Xiaosen Guo, Yong Wang, Kirk E. Lohmueller, Simon Rasmussen, Anders Albrechtsen, Line Skotte, Stinus Lindgreen, Mait Metspalu, Thibaut Jombart, Toomas Kivisild, Weiwei Zhai, Anders Eriksson, Andrea Manica, Ludovic Orlando, Francisco M. De La Vega, Silvana Tridico, Ene Metspalu, Kasper Nielsen, et al. (2011). An Aboriginal Australian Genome Reveals Separate Human Dispersals into Asia. *Science, 334* (6052), 94–98.

Ratzinger, J. (2000). *The Catechism of the Catholic Church.* Vatican Press.

Reas, C. (2005). *Software Art and the Algorithmic Aesthetic.* MIT Press.

Regulation. (EU) 2016/679. General Data Protection Regulation (GDPR).

Rigotti, C. (2020). Sex Robots through Feminist Lenses. *Filosofia, 65,* 21–38.

Ringle, W. M., Gallareta Negrón, T., & Bey, G. J. III. (2008). *The Return of Quetzalcoatl: Evidence for the Spread of a World Religion during the Epiclassic Period.* Cambridge University Press.

Ripple, W. J., Christopher Wolf, Jillian W. Gregg, Johan Rockström, Michael E. Mann, Naomi Oreskes, Timot hy M. Lenton, Stefan Rahmstorf, Thomas M. Newsome, Chi Xu, Jens-Christian Svenning, Cássio Cardoso Pereira, Beverly E. Law and Thomas W. Crowther. (2024). The 2024 State of the Climate Report: Perilous Times on Planet Earth. *Bioscience, 0,* 1–13. https://doi.org/10.1093/biosci/biae087

Riskin, J. (2016). *The Restless Clock: A History of the Centuries-Long Argument over What Makes Living Things Tick.* University of Chicago Press.

Roebroeks, W., & Villa, P. (2011). On the Earliest Evidence for Habitual Use of Fire in Europe. *Proceedings of the National Academy of Sciences of the United States of America (PNAS), 108*(13), 5209–5214.

Rosenberg, D. (2010). *The Meaning of "Man" in the Middle Ages.* Pontifical Institute of Mediaeval Studies.

Rosenberg, E. (2009). *Human Nature: A Philosophical Introduction.* Stanford University Press.

Russell, S., & Norvig, P. (2020). *Artificial Intelligence: A Modern Approach* (4th ed.). Pearson.

Sacks, J. (2003). *The Dignity of Difference: How to Avoid the Clash of Civilizations.* Continuum.

Sacks, O. (1985). *The Man Who Mistook His Wife for a Hat and Other Clinical Tales.* Summit Books.

Said, E. W. (1978). *Orientalism*. Pantheon Books.

Sandel, M. J. (2007). *The Case against Perfection: Ethics in the Age of Genetic Engineering*. Harvard University Press.

Sanders, R. (Director). (2017). *Ghost in the Shell* [Film]. Paramount Pictures; DreamWorks Pictures; Reliance Entertainment.

Sankararaman, S., Swapan Mallick, Nick Patterson & David Reich. (2016). The Combined Landscape of Denisovan and Neanderthal Ancestry in Present-Day Humans. *Current Biology*, 26(9), 1241–1247.

Scharre, P. (2018). *Army of None: Autonomous Weapons and the Future of War*. W.W. Norton & Company.

Schiebinger, L. (2004). *Nature's Body: Gender in the Making of Modern Science*. Rutgers University Press.

Schiff, D. (2023). *Judaism in a Digital Age: An Ancient Tradition Confronts a Transformative Era*. Springer.

Schwab, K. (2016). *The Fourth Industrial Revolution*. Crown Business.

Scott, R. (Director). (1982). *Blade Runner* [Film]. Warner Bros.

Scully, M. (2019). Eugenics and Its Contemporary Echoes: Ethical Considerations in Genetic Engineering. *Journal of Medical Ethics*, 45(2), 122–128.

Searle, J. R. (1980). Minds, Brains, and Programs. *Behavioral and Brain Sciences*, 3(3), 417–457. https://doi.org/10.1017/S0140525X00005756

Seth, S. (2010). *Subject Lessons: The Western Education of Colonial India*. Duke University Press.

Sharkey, N. (2010). Saying 'No!' to Lethal Autonomous Targeting. *Journal of Military Ethics*, 9(4), 369–383.

Shapin, S. (2007). *The Scientific Revolution*. University of Chicago Press.

Shelley, M. (2019). Frankenstein. Wisehouse Classics.

Shin, T., Kraemer D., Pryor J., Liu L., Rugila J., Howe L., Buck S., Murphy K., Lyons L., & Westhusin M. (2002). A Cat Cloned by Nuclear Transplantation. *Nature, 415*(6874), 859. DOI: 10.1038/nature723

Shor, P. W. (1994). Algorithms for Quantum Computation: Discrete Logarithms and Factorization. *Proceedings of the 35th Annual Symposium on Foundations of Computer Science*, 124–134.

Singh, P. (2011). *Sikhism: An Introduction*. I.B. Tauris.

Slon, V., Viola, B., Renaud, G., Gansauge, M.-T., Benazzi, S., Sawyer, S., Hublin, J.-J., Shunkov, M. V., Derevianko, A. P., Kelso, J., Prüfer, K., Meyer, M., & Pääbo, S. (2017). A Fourth Denisovan Individual. *Science Advances, 3*(7), e1700186.

Smith, D. L. (2021). *Making Monsters: The Uncanny Power of Dehumanization*. Harvard University Press.

Smith, L. (2020). *Chimeras in Greek Mythology: Embodying Chaos and Complexity*. Harvard University Press.

Smith, P. (2008). *An Introduction to the Baha'i Faith*. Cambridge University Press.

Softbank Robotics. (2015). *Pepper: The First Social Humanoid Robot*. Retrieved from https://www.softbankrobotics.com

Solms, M. (2015). *The Brain and the Inner World: An Introduction to the Neuroscience of Subjective Experience*. Karnac Books.

Solms, M. (2021). *The Hidden Spring: A Journey to the Neuroscience of the Soul*. W.W. Norton & Company.

Sontag, S. (1977). *On Photography*. Farrar, Straus and Giroux.

Sorkin, A. (2017). Interview with Sophia the Robot: A Conversation at the Future Investment Initiative. *The New York Times*.

Sosis, R., & Alcorta, C. (2003). Signaling, Solidarity, and the Sacred: The Evolution of Religious Behavior. *Evolutionary Anthropology, 12*(6), 264–274.

Sparrow, R. (2014). Ethics, Eugenics, and Politics. In *The Future of Bioethics: International Dialogues* (pp. 139–153). Oxford University Press.

Stout, D. (2011). Stone Toolmaking and the Evolution of Human Culture and Cognition. *Philosophical Transactions of the Royal Society B: Biological Sciences, 366*(1567), 1050–1059.

Stringer, C. (2012). *The Origin of Our Species*. Penguin.

Strubell, E., Ananya Ganesh & Andrew McCallum. (2019). Energy and Policy Considerations for Deep Learning in NLP. *Proceedings of the 57th Annual Meeting of the Association for Computational Linguistics*, pp. 11–16. https://aclanthology.org/P19-1355/

Stump, E. (2003). *Aquinas*. Routledge.

Styx. (1983). Domo Arigato, Mr. Roboto [Song]. On *Kilroy Was Here* [Album]. A&M Records.

Sullivan, N. (2017). *Robot Sex: Social and Ethical Implications*. MIT Press.

Sunstein, C. R. (2001). Republic.Com. Princeton University Press.

Susskind, R., & Susskind, D. (2022). *The Future of The Professions: How Technology Will Transform the Work of Human Experts*. Oxford University Press.

Sutton, R. S., & Barto, A. G. (2018). *Reinforcement Learning: An Introduction*. MIT Press.

SWANA. (2021). *AI in Waste Management*. Solid Waste Association of North America. Retrieved from https://swana.org/News/Blog/Swana-Post/Swana-Blog/2023/12/11/How-Ai-Is-Revolutionizing-Solid-Waste-Management

Tachibana, M.,Sparman, C. Ramsey, H. Ma, H-S. Lee, M. C. T. Penedo, & S. Mitalipov. (2012). Generation of Chimeric Rhesus Monkeys. *Cell, 148*(1–2), 285–295.DOI: 10.1016/j.cell.2011.12.007

Takahashi, K., & Yamanaka, S. (2007). Induction of Pluripotent Stem Cells from Adult Human Fibroblasts by Defined Factors. *Cell, 131*(5), 861–872. https://doi.org/10.1016/J.Cell.2007.11.019

Tatar, M. (2019). *The Hard Facts of the Grimms' Fairy Tales*. Princeton University Press.

Tattersall, I. (1995). *The Fossil Trail: How We Know What We Think We Know about Human Evolution*. Oxford University Press.

Taylor, B. (1999). *The Surrealists: Revolution in the Making*. Tate Publishing.

Tedlock, D. (1985). *Popol Vuh: The Definitive Edition of the Mayan Book of the Dawn of Life and the Glories of Gods and Kings*. Simon & Schuster.

Tegmark, M. (2017). *Life 3.0: Being Human in the Age of Artificial Intelligence*. Alfred A. Knopf.

Terence. (1976). *The Comedies* (S. G. G. Scullard, Trans.). Penguin Classics.

Thompson, A. (2019). *Tattoo: The Anthropology of an Ancient Art*. Reaktion Books.

Tiggemann, M. (2015). The Impact of Social Media on Body Image Concerns in Young Women. *Body Image, 13*, 38–45.

Tilly, C. (1990). *Coercion, Capital, and European States, AD 990–1992*. Basil Blackwell.

Tilly, C. (2005). *Identities, Boundaries, and Social Ties*. Paradigm Publishers.

Todorov, Tzvetan. (1984). *The Conquest of America: The Question of the Other*. Harper & Row.

Tomasello, M. (2001). *The Cultural Origins of Human Cognition*. Harvard University Press.

Tononi, G. (2008). Consciousness as Integrated Information: A Provisional Manifesto. *Biological Bulletin, 215*(3), 216–242.

Tononi, G. (2019). *Phi: A Voyage from the Brain to the Soul*. Pantheon Books.

Tononi, G., & Koch, C. (2015). Consciousness: Here, There and Everywhere? *Philosophical Transactions of the Royal Society B: Biological Sciences, 370*(1668), 20140167.

Treija, S., Bratuškins, U., Koroļova, A., & Lektauers, A. (2021). Smart Governance: An Investigation into Participatory Budgeting Models. *Environmental Sciences Proceedings, 11*(1), 30. https://doi.org/10.3390/Environsciproc2021011030

Tufekci, Z. (2015). *Twitter and Tear Gas: The Power and Fragility of Networked Protest*. Yale University Press.

Turkle, S. (2011). *Alone Together: Why We Expect More from Technology and Less from Each Other* Basic Books.

Turner, V. (1969). *The Ritual Process: Structure and Anti-Structure*. Aldine Publishing Company.

Turner, V. (1986). *The Anthropology of Performance*. PAJ Publications.

Twenge, J. M., Joiner, T. E., Rogers, M. L., & Martin, G. N. (2017). Increases in Depressive Symptoms, Suicide-Related Outcomes, and Suicide Rates among U.S. Adolescents after 2010 and Links to Increased Smartphone Use. *Clinical Psychological Science, 6*(1), 3–17. https://doi.org/10.1177/2167702617723376

UN Environment Programme. (2018). *Global Environment Outlook 6: Healthy Planet,* Healthy People. Cambridge University Press.

UN Office of the High Commissioner for Human Rights. (2024, November 8). *Six-Month Update Report on the Human Rights Situation in Gaza: 1* November 2023 to 30 April 2024. Retrieved from

UNESCO. (2023). Recommendation on the Ethics of Artificial Intelligence. Retrieved from https://www.unesco.org/en/articles/recommendation-ethics-artificial-intelligence

United Nations. (2005). *United Nations Declaration on Human Cloning*. UN General Assembly.

United Nations. (2018). *The Impact of Environmental Pollution on Human Health*. United Nations.

United Nations. (2020a). *The Sustainable Development Goals Report 2020*. United Nations.

United Nations. (2020b). *World Population Prospects 2019*. United Nations.

United Nations Framework Convention on Climate Change. (2019). *The Paris Agreement*. United Nations.

Varela, F. J., & Shear, J. (1999). *The View from Within: First-Person Approaches to the Study of Consciousness*. Imprint Academic.

Verbeek, P.-P. (2005). *What Things Do: Philosophical Reflections on Technology, Agency, And Design*. Pennsylvania State University Press.

Verhoeven, P. (Director). (1987). *RoboCop* [Film]. Orion Pictures.

Vincent, S., & Brackley, J. (Creators). (2015–2018). *Humans* [TV series]. Channel 4, AMC.

Volkswagen. (2019). *Optimizing Traffic Flow with Quantum Computing*. https://www.volkswagen-group.com/en/press-releases/Volkswagen-Optimizes-Traffic-Flow-With-Quantum-Computers-16995.

Vosoughi, S., Roy, D., & Aral, S. (2018). The Spread of True and False News Online. *Science, 359*(6380), 1146–1151. https://doi.org/10.1126/Science.Aap9559

Wacher, J. (Ed.). (2001). *The Roman World*. Routledge.

Wada, K., & Shibata, T. (2011). Robot Therapy: A New Approach for Mental Healthcare of the Elderly – A Mini-Review. *Gerontology, 57*(4), 378–386. https://doi.org/10.1159/000319015

Wajcman, J. (2010). Feminist Theories of Technology. *Cambridge Journal of Economics, 34*, 143–152.

Waldby, C., & Mitchell, R. (2006). *Tissue Economies: Blood, Organs, and Cell Lines in Late Capitalism*. Duke University Press.

WCS. (2020). *New Artificial Intelligence Could Save both Elephant and Human Lives*. https://news.Mongabay.Com/2020/09/New-Artificial-Intelligence-Could-Save-Both-Elephant-And-Human-Lives

Weber, M. (2002). *The Protestant Ethic and the Spirit of Capitalism*. Penguin.

Weymar, D. (2018). Alexa, Play Me Some Art: The Role of Voice-Activated Technology in Contemporary Art. [Art project].

Whiten, A., Hinde, R. A., Laland, K. N., & Stringer, C. B. (2012). *Culture Evolves*. Oxford University Press.

Wilkinson, R. H. (2003). *The Complete Gods and Goddesses of Ancient Egypt*. Thames & Hudson.

Wilmut, I., Schnieke, A. E., McWhir J., Kind, A. J., & Campbell, K. H. S. (1997). Viable Offspring Derived from Fetal and Adult Mammalian Cells. *Nature, 385*(6619), 810–813. https://doi.org/10.1038/385810a0

Wiredu, K. (1996). *Cultural Universals and Particulars: An African Perspective*. Indiana University Press.

Wirtz, B. W., & Müller, W. M. (2019). An Integrated Artificial Intelligence Framework for Public Management. *Public Management Review,* 21(7), 1076–1100. https://doi.org/10.1080/14719037.2018.1549268

WMO. (2021). *State of the Global Climate 2020.* World Meteorological Organization.

Wolf, E. R. (1982). *Europe and the People without History.* University Of California Press.

Wolfe, C. (2009). *What Is Posthumanism?* University Of Minnesota Press.

Wood, B. (2019). *Human Evolution: A Very Short Introduction (Very Short Introductions).* Oxford University Press.

WorldHealthOrganization.(2024a).Ambient(Outdoor)AirPollution.*WorldHealthOrganization.* https://www.who.int/News-Room/Fact-Sheets/Detail/Ambient-(Outdoor)-Air-Quality-And-Health

World Health Organization. (2024b). Artificial Intelligence for Health: Supporting Countries to Deploy Responsible AI Technologies to Accelerate Equitable Health for all. World Health Organization. https://www.who.int/Teams/Digital-Health-And-Innovation/Harnessing-Artificial-Intelligence-For-Health

Wrangham, R. (2009). *Catching Fire: How Cooking Made Us Human.* Basic Books.

Wynter, S. (2003). Unsettling The Coloniality of Being/Power/Truth/Freedom: Toward the Human, after Man, its Overrepresentation—An Argument. *CR: The New Centennial Review,* 3(3), 257–337.

Zaman, K., Marchisio, A., Hanif, M. A., & Shafique, M. (2023). A Survey on Quantum Machine Learning: Current Trends, Challenges, Opportunities, and the Road ahead. *Arxiv Preprint* Arxiv:2310.10315

Zuboff, S. (2019). *The Age of Surveillance Capitalism: The Fight for a Human Future at the New Frontier of Power.* PublicAffairs.

Index